NEON AND CHLOROPHYLL

NEO-MODERN NATIVE LITERATURE
FROM THE INSTITUTE OF AMERICAN INDIAN ARTS

2003-2004

EDITORS
Tristan Ahtone
Ishmael Antar
Jody Barnes
D. G. Okpik
Sara Marie Ortiz
Deleana Otherbull
Penina Partsch
Douglas Two Bulls
Orlando White

For information:

Creative Writing Program
Institute of American Indian Arts
83 Avan Nu Po Road
Santa Fe, New Mexico 87508

This book is available through our distributor:

Small Press Distribution
1341 7th Street
Berkeley, CA 94710
(510) 524-1668

ISBN: 1-881396-20-7

Recent titles from the Institute of American Indian Arts
anthology series include:

Bone Light (2003)
Fourth World Rising (2002)
Maple Vapor (2001)

TABLE OF CONTENTS

POETRY

FICTION

PLAYS

NON-FICTION

CONTRIBUTORS' NOTES

POETRY

STILL LIFE
Britta Andersson

I draw flowers with no name
no seed, no soil.

I call them something inside myself;
the coiled parts in my stomach, beneath my chest.

They bloom out of season
and in the dark.

Shadows burnt at the edges and curved like a fist;
a calyx that lures a tongue.

They thrive and die
in lacquer, in wax.

Flowers with salt veins the scent of black;
a crows caw, hatched in water.

AGAINST LIGHT
Britta Andersson

Explicate the center of my eye:

a double xerox
a triple helix
fragments between the skin
bigger around the calyx
bent along the wrist

Broken bulbs shatter light
before their birth can breathe a shadow.

A red geranium stiffens closed
the eye that stems inside me.

I read the same words
from the opposite side
of the other page

Black blossoms bend backwards

to look inside.

KINETIC SNOW
Britta Andersson

---

xeroxed specks
on dark white fields
    dangle transparent lines
    carve pinched lead eyes into black:

       imprints, tracks

   quick circuit energy
     syntax drums
   inside ear sticks down throats

    blinks   thousands
   breaths   inverted

melting logic   twisting tongues

      magnetic paper:

letters, marrowed   fingers, bones

    panic-struck

      on glass

INTERSECT
Britta Andersson

---

points and lines
exist to part from origin
and bloom between each other
new words that don't exist

read a broken blossom
in black space

nothing
between them

white: a clean and empty shadow
offers little light beneath its skin

negative space
gives depth to painted figures
dispersed from melting tongues
where language seeds in silence

lines and points on fields
break open
blind words on blackened light

MIRRORS
Britta Andersson

Sometimes I look into a mirror and pretend I remember who I am.

The lines on my face grow into one another, connecting each of their births to the center of my eye. It is in that eye that I see the reflection of the window open behind me, tiny shivers of moonlight. An arctic wind rushes the length of my skin.

I used to think it was magic glass, the kind that could translate silent conversations between my real eyes and the marble ones that stared back. But then I grew up, and there was nothing to de-code in an empty face.

A pale blue sliver makes it shape through the blinds, landing on smooth tiles against the wall. It looks like a broken icicle in two places, between them, a frozen star. I want it to melt in my hand, let it bleed down my wrist and chime off the floor, but there is nothing to touch.

Sometimes I would run hot water in the sink, watch steam curl slowly, a murky haze. It was a map to the mute space inside me. I would try to get it to talk, write words on the clouded pane that sounded like colored water. Water takes on the color of the sky or the glass that holds it, changing like a language in deaf ears. In the mirror, the words appear backward and hard to read.

Is it the blackest part that lets in light? My eyes dilate; they talk to each other with nothing to say. Night leaks through the window; shadows escape inside.

The bathroom has two dim bulbs that shine in unison. I turn off the lights—
Too dark for pretending.

ARTICLE
Britta Andersson

My fingers are ink-marked and smeared
by words
masked with red and yellow stains.
The colors blend like ice in plastic cups.
Voices dim quickly.
a woman picked coffee,
still red, and spent
cold nights on concrete; on her back
a son, now the shape of dusk.
was it written on his palm?
Banana leaves scatter with wind
calm sounds and shuffled paper.
I cannot read the small scream
from the mouth of the strange man;
his wide eyes reflect a flash.
I trace the length of the photographer's name.
Letters conjure noise.
My ears push back, close

BUTTERFLY SKIN
Britta Andersson

A nail half bent and tarnished
Reveals the center of a pine knot;
A woman's face carved from shade.

Shadows split from lodge pole and light.
Shrapnel amidst the lilies, jasper under layers of moss.

A cloud shaped rose
Combines the scent of smoke and breeze
Softer than rain on skin.

Calyx singe on the edge of grass.
Crushed needles inside a wrist.
Bent cocoons in open wind.

What is optional to touch?

A pulse inside of stone,
A peak and bending sigh,
A valley in fetal position

Branches exert an opaque breath across the sky.
Butterflies multiply.

Wings are tissue beneath the flesh.

VOLUME ZOOM
Britta Andersson

A living clock
counts numbers, sideways
　　　on her back.

Plastic arms, cotton-spun
pinch water's transparent hour.

Things quiver and wake in glass.

A vein spills glass beads through skin;
　　micro tongues with secret eyes,
　　　　fingers encased with stone

　　black concrete, black holes
　　　black specks under breaking clocks.
Red seconds pass young ticks
　　　through blank sheets
　　　　　and black tocks.

To tell time,
pry confession out of a steel-rimmed egg:

Take a deep breath

Take another deep breath

You may breathe.

SPRING POND

I. M. Antar

At the sun's breaking of horizon
Vernal equinox   is pierced
by a sparrow's darting cry
a lily opens on its pad
          resting
atop the waves of a pebble's ripple
this pond
is a wrinkle leasing time
a soft space   in silken fabric folds
she weaves   the orange
of spring's sunrise
as wind ripples
each lonely   blade   of grass

DUST AND NUMBERS STAINED IN INK
I. M. Antar

Second power   small two
raised above a variable: x
at half a #'s height
dusted chalk board   grinds steel
against my eardrums
filled with #'s which to me are a false inference
their lies highlighted   by a purple suede sweater

a wrinkle catching light
multiply the suede
by the wrinkle
and rent a clock
then get a hammer just to kill some time
stained satin ink
cubed
the third power of ink
is light refracting the grains of a sheet of paper:
dust

dust stings the eyes because it is jaded
a purple bag holds one pound of dust
          becoming my cynical friend
with as much hope as a brass cymbal
turned upside down   to catch rain water

flourescent captivity
pounding my eyes into submission
until they swell and burst
like a red rubber water balloon attached to a running faucet

the purple suede sweater
stained with red ink

from a dry erase marker
as the board loses its footing and crashes
into the suede sweater, dust covers the ink stain
and the flourescent light begs
the ink   to shimmer
under its pulsating
illumination

INK TRACED
I. M. Antar

Scratching the surface of a skull
shrouded
in a paper-mache mask
with the tip of my fountain pen

I can smell the ink cascading from the tares
made by my pen: the smell of desperate autumn ashes
hanging on to the crisp wind

a plastic skull filled with ink
 – sitting on the dash of his '89 Honda Civic –
tosses it's own blood out of the window at 60 miles per hour
the liquid splashes across asphalt   over the 1800 miles of desert
to the plains   across cement bridges stretching over rivers
         and canyons

The ink marks a path over this continent
         in the desert
it boils away
         in the plains
it takes its time to soak into the black tar
it drips over mountains
and hangs from blue-shaded granite walls
         that hold the sea at bay

Hush little baby
Hush little baby

would you   *Pull My Daisy?*
I want a yes or a no, don't give me a maybe
and...

Decapitate the dandelion!
Decapitate the dandelion!

Because it's just what we need
expel a wicked breath, come on now spread the seed

and... a desperate allusion to *Visions of Cody*
or White Horse Tavern, corner echos "Blow me"
the sound a reflection thats been all bounced around
3 decades or 4... cause... it can't find the door
and... in that corner its Ginsberg's voice begging for pleasure
Then Corso came round with a yardstick to measure
but..
Would you *care* to *whack* me over the head with my
            mexico blues?
I've lived with *Dharma Bums* in bowling shoes and stoners who
            actually read the news

at San Remo's I extracted the frequency of granite laced with tiger's
            eye
                        from the sound of a muted Miles trumpet sigh
                        Gregory was born right down the street, 'bout 40
                        yards up an ally
                        10 dragon flies at his feet
                        the fizz of a humming dragon with a cleft lip
                        drips into his ears

all peach in flavor and fuzz, later it's oil for the gears
so that he can sing *The Happy Birthday of Death*, and *The Vestal
Lady*
cause oh she be commin' and *Howl*-ing to the night as she does

after poetic incarceration till 1950
its poetic masturbation oh yes it tickles me
spiral speed the bullets spin
he was a subterranean
now hush little baby
       hush oh hush my little daisy

       I promise not to pull on you too hard

INDIAN SWIRL
Leslie Gee

Mutt.
Halfbreed.
Another Oklahoma Indin' swirl.

Grew up runnin' wild--
Red dirt under my nails,
Horny toads in my pockets,
Green sticker covered shoelaces,
Howlin' at the half moon with the coyotes at night,
Fearin' the wrath of God in the Indian Bible belt.

Became the beloved project of my Irish-Catholic step-dad,
Hell bent on civilizing this Indian.
Spoon-fed me musicals and literature, art and theatre,
Cut Vonnegut and Hemmingway into bite size pieces on my plate.

Now I dance the dirty boogie back and forth across the line,
Kick dirt on both sides and love it.
And hate it.

MAN O MAN. IF ONLY I HAD…
Leslie Gee

A sweet raspy sangin' voice,
A twenty-inch waist
& Three tall black back up singers
That followed me everywhere.

Me and my boys,
We'd strut down town,
Doin' thangs, lookin' good.

Each would stand
Six feet tall and lean
With a glorious afro and
Gleaming white teeth.

They'd be funky,
Not your average matching suites-
More like Sly and the Family Stone,
With platforms and boas.
They'd gyrate and snap.

Like Mary and Gladys and James,
My own entourage to back me up,
Musify every word I spoke.

Please oh please. The girls wants fries.
Shoo bee doo wap. Give mamma some fries.

She don't feel like takin' out the trash.
Na na na no no. She ain't takin out that trash.

Somebody somebody somebody betta
Have that catbox clean by the time she walks in the door.

3221 GERALDINE
Leslie Gee

Our tummies full of dollar beers
And cheap tequila,
We stole a six-pack and a bottle of wine
From my mother's new kitchen,
Hijacked the U-Haul late one night.

I don't remember the ride there,
Pulling into the crooked driveway,
Flipping the For Sale sign out front
A hard, mean bird,
The long fumble with my keys,
The first whiff of air,
That carried only a trace of our old smell.

We went out back and got some firewood.
Three Indians, we still cheated,
Blasting the gas at first, letting the kindling catch,
Then turning it off.
And sat down on the floor in front of it.

I began to reclaim,
Show my comrades
My old stomping grounds.

The room where Bruce Lee's spirit lived,
The vent the squirrel crawled out of on Christmas Day,
The empty bar that used to house my parents stash,
The piano's old spot where I played a mean right hand,
The makeshift theatre, and the closet
Where they hid the Chewy Chewbaca mask.

And my room.
I flipped on my old light switch
And noticed several footprints in the carpet.
Buyer prints.

I stomped out.
Said my goodbyes to the
Empty bedrooms my siblings occupied,

To the big pink bathroom,
Where I used to jump up and down,
Then look quick in the mirror
And watch my butt jiggle,
To my parents big ugly room,
The un-matching tile in the kitchen
And dining rooms,
To the conservatory and the back yard.

And sat back down in front of the fire
I nestled down deep into the
thick gray carpet for a fireside nap.

When I woke, the flames of the fire I lit
Were still dancing high.
Through the glass French doors,
The stars were still shining.
The boys were still sitting beside me,
In front of the warm flames drinking.

We got up,
Gathered our empties,
And left that fire burning.

THE FUGITIVE (after Caravaggio)
Javier Gonzalez

I run stepping over corpses, sliding on blood and excrement. They follow me and say: this man has not been seduced by the celestial choirs, prophet's visions or the crisp tongue of muses! You see, an eye is not a perfect sphere pinked by capillary webs; it is a liar and a judge who trap us outside of ourselves. The eyelids are hatching the egg, one day a bird that will learn how to fly. You see, circles are hunchbacks; squares resemble Siamese twins, triangles, bow-legged. I learned how to lie with a golden mean, I tell the truth with a crooked branch through fields of burning stakes. I run barefooted over broken glasses, they follow. I am a fugitive who concaved the surface of Cartesian planes, filling rooms with earth, disinterred bodies licked by light, like cotton balls soaked in iodine. I run across empty parks, over roof tops, jumping the walls, breaking through windows. My sentence hangs from the mouth of a barrel. I fill my pockets with stones arming my self with a nail, their dogs are sniffing my trail, I hear them yawning. I live under the cover of my own shadow, picking from the crusts, milk skin pinched between fingers. Inside the darkened rooms, a moribund candle light bites my eyeballs like maggots gnawing an apple. In the shadows my faceless expression is a rumor of revolting fire-flies. With a torch of nocturnal luminescence I see: black flowers in black vases, black papers over a black table, black windows overlooking black landscapes. The skin of darkness is a silk rag. When the cloth slides and falls from my shoulders, I find my self cut away from the space, swimming in a pool of liquid onyx, staring with saltpeter and nitro eyes, a promise of green leaves after my death of every winter.

LEFTOVERS
Javier Gonzalez

Fleas are crueler at breakfast,
stabbing little tongues that miss the heart.
Connoisseurs of vintage lymph,
they slurp pores like lemon-tea,
burying their lost teeth under wrinkles.

Fleas are amazing:
unlike humans,
they don't banquet at open wounds.

I learned this by asking a beggar,
an erudite versed in the area.
To find him,
I followed a trail of bleeding footsteps.

If the shoeless soles of his feet
could speak,
they would describe the facial complexion
of this town.

He was standing on the corner he calls home,
like a patch mending the ragged wall.

Ate a lobster yesterday,
but today, he is again a tramp,
and if he threw away all his money
it wouldn't make him any poorer.

He's got big hands, a true gift he says,
the envy of other mendicants, even though
he is just an infant who eats boxes full of
oranges in his dreams,
and the list of the things he lacks:
bread by bread, brick by brick,
line by line, stretches into three volumes.

How many coins, I ask, could
A flea infested-child have
if he doesn't know how to count?

He didn't answer, kept scratching the flea's bite
in the place that connects the body
with its shadow,
thinking that the fleas
will not leave any crumbs
tonight.

PASSING A STRING THROUGH THE HEART
Javier Gonzalez

One remembers
simple things
made out of wood
or wheat,
like wrinkled silk hands rolling the thumbs,
kids running through the fields
holding the carcass of a little bird,
the voice of a friend saying nothing,
the taste of raindrops,
a lover's touch in the dark,
the smell of a tree that knew
the harvest bug song,
a tore wallpaper with butterfly prints,
the butter-lime color of the eastern hills
at sundown,
a fossil that held the door open.
One remembers stories
Like the one of a river
that stole the most beautiful girls of a town,
or the story of a man who died
when he found himself,
or my neighbor,
who taught Spanish to a parrot
to keep her company.
And one remembers too,
That we are also stories
remembering things that live
despite of us
waiting for the joy
of tomorrow's light,
the light of the sun
that will keep us warm
as if it were
remembering us,
passing a string through the heart.

CHEESE
Javier Gonzalez

Silence = paper swans swimming on glass.
Thirst =Vesseltongue, vocal chords extracted from their cavity.
White horse + cymbals + microphone = sugar cubes dissolving in
snow.
Bent knee + a board tied to the shoe = ectoplasm.
1, 3, 4 hands inside the bucket = watch your back! they are around
the corner.
Ghost = hello.
Body fat = kill me so I can feel.
Interview + neon sign =59 cents.

Rose=eternity = frozenface waiting for its turn to speak.

Leaf + motor oil + ribbon in gold letters + alabaster = vertical line
smiling in its way up.

I used to walk with a Greek sculpture's nose inside my pocket but my
Brazilian friend thought this was bizarre. She talked about Sao Paulo as
if it were a cloud of smoke. They had the metro and that cheap
melodrama "Central station" = what did we have instead?

Screwdriver + can + grease = child 1988.
Noise eraser + nose = order-1.
Boots + effigy + cracked floor x barrel x n2 – gray + cat = arms
sticking out.
Round thing + wall = almost emptiness, try again later.

Art was dead=smelled like cheese + mold.

Coat + crystal + math = person in the background singing.
Dryfishupsidedown = wound. Wound = headless man buys a comb.
Stone + telephone = wrong number.
Lantern – blanket = club soda.

Empty box =
chocolate taste in mouth =
reflexes delayed =

Babylon =
Marie Louis Pegarie =
itching humanity =
Pegasus =
orange =
brown ink =
iron =
name =
tree =
?

she was 1/4 African, 1/4 Indian, 1/4 Portuguese, 1/4 vegetarian= she was a "boutique hippie" and used to spit on the street.

=sled+ the fall of Rome +horse kick in the face and laughing + thumbnail shaped box + an unforgettable night! +monkey mask + money= art.

Life=death=life= do not enter, exit only, use another door please.

TONGUES TURN INTO EARS
Javier Gonzalez

I cut off my tongue and
planted it deep,
one day it will be a tree of ears.
Wet rooted tongues
inserting black seeds
inside a concentric circle of ears.
They grow, like skinny
fingers scratching the earth,
stringing knots around stones,
opening roads of swirling muscles.
Grow, entwined with bones of clay,
thirsty for calcium and amethyst,
whistling threads into blackness,
tickling minutes of iron,
piercing the silence, the milk
from uterine icebergs,
mapping flanks with strands of slime,
webbing entrails, the tongues
grow mending their trails.
A tongue stuck out from the earth
is the umbilical chord or my ears
listening to the wind.
My ears are blanketed with moss,
I see people come out of a common grave:
they open it and fill it with sugar.
Around the tree trunk,
the dance begins
when the bodies touch
each other's fingers
of glass glimmering breath.
Their tongues sweat
grains of rice, corn, salt.
Their ears talk:
Tongues, are not knives!
Ears, are not scars!
Right feet in,

left feet back.
Hands to their waists,
making circles with their hips.

THE FLY
Javier Gonzales

Thrown on the ground of a wasteland
thoughts are broken chalk lines
over white shell walls of lime.
In an invisible corner of the screen
a dead fly is a threat
placed by divine intervention
or by the wind
onto my cranium's ceiling.
I scratched it and scratched it
scratched it and scratched it
scratched it and scratched it
until the stucco bleeds the milk
but the dead fly was there still,
staring to me in silence
I could almost hear it saying
"never! – never more!"
So I looked at it closer,
but the fly was gone
and in its place,
a little hole instead,
peeping through it
I saw
a tiny light out of focus
somewhere on the other side.
Suddenly
the hole started to grow
quickly, unannounced as a third world quake
the cracks covered the roof of my mind
like the web of a giant spider
from one side to the other.
The ceiling came down in a snap
and I
was dazed for a while,
but now,
I can see the stars.

I stare at the bottom of the cup with dry eyes, people come and go, I don't know names, I never recognize a criminal by his face. And who are the criminals? Those who know what they are doing? Those who feel they are like gods when the skin they pinch becomes red with blood? People come and go in the wheel of life. What happens to missing people when we stop looking for them? Are they still missing? Some mothers won't have peace until their children are buried, some people bury their treasures, and others bury their garbage, some bury to forget, others to remind. People come and go, asking for the receipt, making sure the transactions are clear, all the pennies counted in ink. It is always important to get the right change, so we can smile, give thanks and forget easily when we turn our backs. Human history is written on the back of a penny. Armies fought, ships sailed, empires crumbled, tears were shed by this penny. So poor is this penny, it costs so much to have it. It feels like something rather precious could have had its place. But we walk, we come and go, we displace water from a tub.

INVENTORY
Javier Gonzalez

I have paintbrushes
to open doors,
to clean frozen ducts
and level wobbling chairs.
Paintbrushes tied to lanterns,
to forks, to fingers, to other paintbrushes.

I have paintbrushes that lost their heads,
I tried to fix them with duck tape.

Paintbrushes with camel hair, dolphin hair
rattlesnake hair, bluebird hair.
Paintbrushes with the hair stuck together
like fists, white haired, with afros, flaming heads,
wounded, in a wheelchair,
veterans from the war of a thousand days.
Paintbrushes with vertebras of doves.

I have immortal paintbrushes
looking like ghosts, although they can be also
the reincarnation of pens.

I have paintbrushes that wear ties, go to their desks
do only what they are told to do
and bend over ledger books.
Paintbrushes that are never home,
leaving menacing messages on an answering machine.
Paintbrushes that give birth every two weeks,
to baby paintbrushes crying to be fed.

I have paintbrushes that think
life is the color of water.

I have paint brushes longing for the day when
they will meet the palette of their dreams,
where there will be enough colors to paint
all the things they have seen.

BARE BLIND COUPLE
Javier Gonzalez

The waxed palms of the angel
began to sweat pearls
when he regained
his acrophobia.

Fog was his beloved country,
the fields, the streets.
Fog was his house, table,
food, bed and blanket.

He appeared in front of her,
naked like a hand
coming out of a pocket,
made visible like clear satin
by the bareness it cloaks.
That's how she would have perceived him
if she could only see.

After traveling a century he stood
an instant
on the tracks of rubber soles,
before caressing
her frigid shoulders,
anticipating that thin flesh
like a film projection, an illusion
would turn to dust with a touch
of his angelic fingertips.

Creaking blind fingers
brushed her torso once and
she rolled her eyes back in a wink
as if her dry pupils could fondle him.

While inspired by the wind
or the flutter of wings,
the sightless girl imagined apricots
shaped like juicy breasts.

Then, the angel kissed
her amber forehead.
Would she still think
that he was the breeze?

BOLERO
Javier Gonzalez

Buried in an underworld tavern
we listened to a bolero
while drinking tears, goodbyes and maybes
like discolored and tasteless margaritas.

Music unfixed our faces from the skulls;
our flesh was running water.

Leaking skin, blood and muscles dripped from our seats into a
        puddle;
trumpets sculpted the anatomy of a rose inside your ears,
our eyes became smoke filling the ceiling's dome.

You asked me again what I was thinking,
and that one time I wished I had an answer.

Looking at each other's faceless, shifting expressions
there was the unbearable sharpness of our future.
And what was our destiny but that tissue ripping sound of out of
        tune guitars?
Eating us from the inside out,
leaving us dry, sucking us into patterns
of school girl uniforms,
as the trio continued to bounce the congas
through "el infiernito"
and the waitress from hell served more "aguardiente"
that we sipped without lips.

You carved your name on the table
offering me your wound in the shape of a tangerine,
the peels of acrid smell brought tears to our sockets,
the seeds filled wooden cracks on the floor,
and no one noticed.

There was this lady dancing with a knife stabbed on her back,
a drunk man removed his scalp and fed his brain to leeches.
The couples on the dance floor were like live rabbits inside a
        cloth bag,
the smell of their sweat drove me to the edge.

BOLERO

One of the singers drowned himself in the bathtub,
the line to the restroom grew longer,
the other two kept singing.
The song was about martyrs pregnant with sorrows
and broken bottles in the head of despair.
I don't remember the rest.

We took a picture of our bones with a cloud's backdrop,
before we looked around
pretending that it never happened.

And then you left or I left first,
who really cares?
But years later
I still wake up with your smell.

TRIFLES
Javier Gonzalez

Human blood is braided rocking in a wooden loom,
the strands of ruby woolen tangles drip
as they are woven blow by blow
by boney artisan's hands into the warp.

Every second day, I try to live skipping days,
writing over x-ray photos,
listening to soccer games through the static.
I have to put hundreds of ghosts to rest,
tying a knot every time I forget,
I can't forget to tie the knots.

Dust storm howls
stirring papers,
tearing sunshades on the market.
Earth adheres to the meat.
We run to take cover,
multitude of feet and legs
exiting the vision range
of dead fish eyes.
A kite flies away,
the carrousel lights up.

Children desire,
pulling their mother's skirts
pointing at trifles at the fair:
plastic revolvers and crosses
pink and blue.
Death is a trifle,
bullets, hard candy.

The kite was made with newspapers
and a rag.

a thread of my sweater was caught
in a half hammered nail,
undoes itself.
I have been walking for miles.

GYUGCUOL
Javier Gonzalez

This poem could have been a butterfly,
the paper stained with cerulean watermarks
these lines, pulsing veins.

This poem could have been a mirror,
here is your image: you are a particle.

This poem is a miracle
although,
it wasn't the most auspicious.

This poem was doomed
before it's beginning,
should not have been written,
it is a poet's cross.

This poem was born poor,
grew up poor,
will die poor.

This poem is made out of hay,
there's nothing in it,
not a needle.

This poem could have dreams and hopes,
meant for solitary readers
soaring inside a shut room.

This poem could be a plane ticket,
to go back somewhere,
or a window and a chair
to see the top of trees.

This poem could have been
anything, not everything
in another flip of coins.

MARANA
Javier Gonzalez

The wayfarer drifts away with tangled roads pumping inside his pulse
highways are dusts revolting at his feet.
                              to the left and to the right
branched to the infinite in arms and strings, in fingers and snakes.
Blurred faces, signs and riddles
spinning like slot machines at light speed
He has already died somewhere along the way
he has lived also or at least that's what he remembers.
New roads bring their crossroads
for each thorn on the sole of his feet
each crossroad brings a labyrinth
                              connecting here and there
                                    before and after
                                    now and then
                                    time and space
                                    point A with point B.
The bum opens his arms up to the sky asking for a sign as roots
                              crawling underground for water
but the constellations are
panting roads ahead of him.
                        Oozing the distance of his path
                              wanting to drink all the rivers in one shot
                              he's thinking where to go next
                              but the road chooses for him
                              and he is already gone
                              he's gone tumbling.
The road pushes and throws him
running faster than he does
walking its own way
breathless he wants to stay still on the bloated and blistered road
in one place and one instant
falling in a dream for a million years
                              dreaming to become the road
                              and the road dreams to be him.

CONFESSION
Javier Gonzalez

Yes, I was utterly wrong,
I thought that humans were vertical wounds
against the horizon, feeding their own fissures
with wood and charcoal.
Knocking constellations with empty heads,
smiling at desire with a missing golden tooth.
And they aren't like that,
instead, humans are just humans
like the songs that birds sing when
braiding with clouds the wind's hair.
Yes, I bred raving conjectures all this years.
I believed,
that stones were fossils,
that hands were flowers
and eyes, hungry wolves.
But, it is not like that,
stones are overlooked tears
hands are smoked glasses,
eyes are roads that always lead to the sea.
Yes, I have never known much.
I thought that liquid meant solid
that three meant two and a half,
that You meant I.
And instead,
Liquid is almost solid,
three means the infinite,
We are the Others not knowing that They are Us.

Fire throbbing flesh, bones of sugar and ice, light cradled inside them. Arms and legs move like those of a hurting puppet hanging from a cloud. Hollow face of a mossy Thursday, night hunger pissing on the door, the mind fades thinking itself, there is earth over the bed. Eyes rolling over rust, wishing there was something to wish for, crossing out words inside the head: wounded by the systems of the universe losing at its own game.

Dream of a garden where no one seems to be dying.

INVENTORY
Javier Gonzalez

I have paintbrushes
to open doors,
to clean frozen ducts
and level wobbling chairs.
Paintbrushes tied to lanterns,
to forks, to fingers, to other paintbrushes.

I have paintbrushes that lost their heads,
I tried to fix them with duck tape.

Paintbrushes with camel hair, dolphin hair
rattlesnake hair, bluebird hair.
Paintbrushes with the hair stuck together
like fists, white haired, with afros, flamingheads,
wounded, in a wheelchair,
veterans from the war of a thousand days.
Paintbrushes with vertebras of doves.

I have immortal paintbrushes
looking like ghosts, although they can be also
the reincarnation of pens.

I have paintbrushes that wear ties, go to their desks
do only what they are told to do
and bend over ledger books.
Paintbrushes that are never home,
leaving menacing messages on an answering machine.
Paintbrushes that give birth every two weeks,
to baby paintbrushes crying to be fed.

I have paintbrushes that think
life is the color of water.

I have paintbrushes longing for the day when
they will meet the palette of their dreams,
where there will be enough colors to paint
all the things they have seen.

SEEDS
Javier Gonzalez

Our wet tongues shine black
when we are silent.
Shredded threads
spiraling in the air
toward the inside of our mouths.
We see petals,
its humid breath
is a white dot in the pupils.
A flower that does not exist
with sharp white edges
against the night,
may grow
through a knot
or a crown
from the crack on the floor.
Upstairs, beds rattle.
Our gazes weave a tree
across the glass window.
A raindrop precedes the hailstorm,
we are this cold.

UNDERWATER DEN
Javier Gonzalez

1

Doves nest under the roof of thundering zinc.
Rain leaks through resin, sand, red hollow bricks.
The facade had once been blue,
inlaid with aquamarines.

2

Feet chill on bare heels,
up a staircase of drift wood and staggering nails.
Moon burns the silhouette, while on the tip of toes,
you disappear behind a slamming door. I have the key.

3

A letter shines inside my pocket, four folding your name,
a flower drawn with an angel's hair
and golden ink over a pentagram.
In another pocket: the pen, blood, the wounded finger.

4

Living on the waving entrails of the present,
inside white petals of warm bed's bosoms, dense the night blossomed.
Hair tied to hair on the pillow, rocking the sifting sand at the bottom
of the sea.
Lovers drowned in a coralline dream, shipwrecked on the white sands
of illusions.

5

Fire, from our den filled
with smoke thick as water,
glass melting, red irons shouting out from charcoal walls.
No rush, we are embracing: tongue to lips, teeth biting teeth.

6

Fireman in black
scoops out remains,
with loaded spooning spades.
In a rain of sapphires, your blazing ashes fall upon mine.

NO THOUGHT
Javier Gonzalez

Purple noon cups the waves
of thick storm water mounds.

Fleshy robes entangle
an ocean of pelvic bones
tired of breeding.

Air bubble inside the brain
threatens to break the head nest,
leaving a salty hair taste
that cannot be lifted by the tongue.

A boy cuts his face away, becomes man,

Beercanmouth spits
the morning news.
Dog translates with
human voice: "the flesh holds no thought"

Fishing pole rests
between blond toes off the coast of Panama.

The day like the boat
has run out of fuel.

LOVE POEM
Javier Gonzalez

Smell of mud,
incense of her sex.
Her dresses hang
like silk ghosts,
her hurting shoes
worn from walking
on imaginary towns.
I scrape her dry lips from letters,
touching fingerprints in a mirror.
I lay my eyes across roads
of red and yellow clay.
Unearthing her tracks,
concavity of her body from a pit
Around the city
I keep on sniffing traces
of her scent.

1982
Javier Gonzalez

Praying mantis fingers
drum across the fallen city.
Muted bodies,
stun gunned, flayed are sanctified
by the removal of genitals.

     Fists to faces, feet to breasts:
the torture chambers are full,
the houses don't hold any whispers.

Spreading pamphlets as dandelion's seeds
     "long live death"
the criminal covered with medals
falls like dead leaves without breeze.
     The city runs out of breath.

Ashes paint wrecked buildings,
pistils spill crosses over the ruins.
     Starvation is our daily bread.

Gray clouds of mosquitoes,
buzzing Mozart, bring the plague.
     We take a beating on the ears.

Sewers are open
like the wrists of a suicide.
     The smell of decay hangs from a hook.

Rolling counter clockwise, pebbles
encircle a black hole at five o'clock.
     We hold on to the names
as if they really existed.
     We hold on to life,
as if it were more that this electric wire.

Sky, vermillion blood.
Eye of earth rains over oceans,
sticking out salty-tongue.

We dig our own hole,
with our arms elongated
by poles.

Far away a lonely mountain
is washed with snow
by one brushstroke.

LA BRUJA
Javier Gonzalez

1
The sweet sap of a malefic curse
congealed carmine creek on lips.
The iron goblet breaks , not a drop spilled.
My eyes turned to see the warnings signs:
skeletons of dew shivering on a glass door.

2
Sweat over white stone,
my opal reflection carries the witch upon shoulders.
I am the beast; she is the master.
Carves with cold crooked knife
her name on my jugular.

3
The jaws of night are open wide, inside, a field of charred trees:
                    Black and red,
              pile of ashes and barb wire mazes,
         rainstorms of DDT and sergeant orange,
children with two heads and no arms.

4
My acquaintances lay dead on my own bed,
I have no place to die,
I search for a hole, crawling,
but all the tombs are taken:
                    Rotten cadavers are rooted
                    like fertilizing a future forest.
I search for a friend,
         for a rope to tie me down to the surface,
         for stones and sticks to hold on to.

5
A rattle calls the wind inside the jungle, the malicious bitch dismounts
         from my back,
crying,   laughing,   and crying again,
while running through tree branches like a scared spider-monkey.

6
Sacred Woman reaches
followed by black jaguars.
White macaws fly overhead.
Half of her face is young like an ageless girl   the other half
        wrinkled, revealing the skull.
One eye is a clear emerald cut in hexagonal facets.   Light
        disappears into her hollow-socket.
She kisses my lips,
jade inside mouth
My thorns transformed into roses. A rainbow bridges over dark clouds.

7
Womb,
inlaid with human-bone-phalanges,
gives birth moribund soul
into a world
of blue over green
rhomboids.

FORTUNE COOKIES AFTER HIRONYMUS BOSCH (1450-1516)
Javier Gonzalez

1. Waiting in line, child drops an egg; her brother licks photos from a magazine.

2. Man in flames running, rush hour.

3. Walking around, writing poems, my shoes are covered with excrement.

4. Don't breathe, the air can kill you.

5. Oh fire, have mercy as we feed you with our flesh.

6. I ask you Earth: keep us warm under your skin.

7. Crystalline water, giver of life, your heart has been poisoned.

8. The devil's bones are fragile as porcelain.

9. We beg, do not cut off our power lines, honorable senators.

10. Iron bird, dropping shells, so precise.

11. Pig face, how many have you murdered today?

12. Ballerina spinning over the burning ruins, would I write these words if I could levitate?

13. A car explodes, windows fall from buildings, I kept playing the violin.

14. Our leaders have everything under control, sacrifice.

15. If they ask me what I think, I will lie if I have to.

16. Inventors, have you thought of an umbrella for the toxic rain?

17. Cleaner, whiter teeth. Hunger

18. Contemplation, staring at the white wall, my mind runs on three feet.

19. To create beauty one needs to rip off the flesh and fill it in with silicone.

20. When cutting a loaf of bread, don't forget to hold the knife by its handle.

21. Pillow is a treasure in a bed where three are sleeping.

22. Girl smiles, crushing a bug between her teeth.

23. New generations, shiny shoes stepping on a puddle of mud.

24. Snow falls on my bed, its sound is between us.

25. Fluorescent light blinks overhead, my thoughts are bar codes; my brain scans them.

DREAM OF A SPECIE
Javier Gonzalez

Inside an eggshell
     substance gains flesh,
an injection of Jell-O
pumps through the quill of the planets
continuing their journey
along three black inches
before taping the corner.

Our beds wobble over waves
closing inside a bubble,
turning bones in to ladders
to reach the candle.

The outer-skin of the collective dream
breaks into squares,
like a massive dwelling complex,
where the bodies talk in their sleep.

The buzz of all tongues
converges in an elbow
mixed with television frequencies
and microwaves.
Our messages are fog bites,
phone conversations
between dead points,
mimes performing
in front of a blind crowd.

Wet
eight fingered
hands
grab our heart
making it jump
out of its niche.

Our sight is unfit
to join up
what we've seen
the night before.

In transit between
the mollusk and
the flying mammal,
we go unnoticed.

When we wake up oblivious,
a spit of blood
is the only record kept
after babbling all night long.

SHIWI WOMAN
Alicia Natewa

During summer feast,
    woman dressed in
        black woven cloth,
            red and green yarn
                across.

        Wearing
           turquoise necklace
              and earrings.
                White moccasin
                  around her legs.

        Men singing
           in the distance.
              Traditional songs
              being sung
           with the beats of
        a drum.

Carry pottery
    blows filled
        with steaming
           red chili stew,
              fresh baked over bread,
                cupcakes, oranges, apples,
                  and other great foods
                    to eat.
      Gave it away to
           the men.
           A prayer
              is said,
              giving
           long life and
        luck to the
    woman.

                    This is who
                      we are,
                        a: shiwi.

[Shiwi: Zuni, A:shiwi: Zuni People]

HOME
Alicia Natewa

I sit on
    the edge of
        Dowa Yalannie.

                I listen
            to the voices
        in the wind
            telling a story
                of long ago.

Colowesi came.
    His great flood upon us.

        Men, women, and children

                cry out

                "Hom ansadu!"

Shiwanni's
    children,
        virgin twins, a boy and a girl
    dewasuk' ya and ladu
to colowesi.

           White lines.
        Twins turn
           to harden stones,
                as the water drains

                slowly.

Dowa Yalannie: Corn Mountain, Colowesi: Sea Serpent, Hom
Ansadu: Help me, Shiwanni: Traditional leader, Dewasuk'ya ladu:
Pray and plant prayer sticks

CERTAIN OBLIGATIONS
Jamie Natonabah

Pull apart eyelids,
  surface;
    white glass.

Pour ink
   into the sockets;
watch the shape of words
  slip
   off the cheek.

The sound of water
   absorbing into wood.

A strand of lint
   caught in an eyelash.
Melting between breath,
   shifting
  into someone's shadow.

Reflections crack
   in the mouth.
Spit throat
    into the dirt.

Drawing bones on skin.

OPENING THE ENVELOPE
Jamie Natonabah

The sound of two people
    behind smoked glass;
the shape of a fly
      crawling on the rim.

White yarn
      braided around wrists.
There is breath
      in words that they share.

Secret letters
      written along the shoulders.
The book that began
      with nine blank pages.
Lines within leaves
      ache before the first snow.

Open palms
      the color of sore throats,
    burned
          into the lower back.
Ice melting between fingers:
      blue ink does not spill.

Memorize roads like circulation.

A FEW SUGGESTIONS
Jamie Natonabah

To breathe;

Open skin
        with bits of broken shell.
It's the sound of birds
                through someone's lips.

Splitting.

To listen;

Crawl over ice,
                press wet lips to the surface:
        black bees burst in closed hands.

Speak inside,
        break bones in the other room.
Half a shadow,
        folded under each eyelid.

To touch;

Remember the minute the wind curled
        and blinded the ear.

Absorb reflections into palms,
        collect dust in deep pockets.

Feel the movement of breath over smooth stone.

To see;

Find the moments when another heart
        stops in it's ninth beat.

Pour mercury into veins,
        portrait on skin is infected.

Watch:
        a strand of hair outlines the moon,
                wire pulled tight around ankles;
        bark stripped from cedar.

Close.

The oil of mango skin
            coats
        the lining of the mouth.

Imagine the air never moved again.

HEART UPON THE EAR
Jamie Natonabah

Caught in the moment
       before sleep,
  warm feathers
spread
        across the back.

The sounds
      of birds at night,
  a heartbeat so clear
       their singing
could not be saved
          for morning.

A wing
      in cupped hands,
  bones as thin as cobwebs.

Heat along the spine,
      burst inside knuckles.

A pulse at the center
      of a palm
            listens
      through four ribs.

Words begin to sour
      in the mouth.

Never press lips against mirrors.

1.

Two shadows
        sewn together;
bones in the legs
                unlace.

Lips heavy
        with sleep.

Black hair
        wrapped
    around a throat.
Fingerprints
        placed   within teeth.

Skin
        coated with glass.

2.

Open
                let the iris leak out.
The shape
        of a bottle
                crushed into grass.

3.

Peel back ribs
        and expose two hearts.
The bodies
                drown
        beside each other.

I trace
        the map of his hands.

65

A VIOLIN IN BLUE
Dg Nanouk Okpik

For Audrey

Morning song: *Aubade*
a black beetle runs across white stone,
intermezzo, allegro,
spring water coda: *passage*

Processional

A knotted pine
leaks sap. It hardens--
then falls. *Interlude.*

In Russia the time is thirty-seven
past the hour. Sixty miles east
a day and thirty-seven minutes past the hour.
Antedate: opus chime.

Burning the coarsely ground myrrh
then surmise unspoiled
wafers of wheat.

Offeratory: Rhapsody

Peeling, curling white birch bark
our flax in linen skin
a vat of saffron to bathe
bright, *grace note in B.*

Recessional: Absolute solo

This fern curls and drinks water next to Chena River.
    I engrave drill bows and polish with cotton circles.

Over there the black whales arch and span,
    four-sided sabers guard the processing barge.

Pollen lands where the air is good. Dig for chert bone.
    Find an antler. Reel in the velvet then map trade.

The small wooden faces flat with skin lined splinters ask:
    Should we prune more trees for pluck or tag and replant?

We the Red Stone people keep our millwork central.
    In the New Stone Age, don't let the paddle wheel rust.

Tie the knitted musk oxen hat with ivory toggles
    firm and fixed, around our chins--

Kiln powder in beveled pools, on beetle rust greens.
    The talc settles, no rain in seventeen days.

Cross puddle mud with dry ankles on earth,
    birth to a metal egg from sledge moss.

Invent a fan to blow the north wind to cool the ivory bone etch.
    The tall grass calls bent birch snowshoes to make tracks.

Do we run a tap dry of soot and sludge to forge roots?
    How many drink wild tea and dip blubber in seal oil?

From the horizon we watch fire opals come from molten rain,
    the clay mass returns to full grass baskets.

ODE TO SALVADOR DALI: OIL ON CANVAS
Dg Nanouk Okpik

I.

A womb of moss and sparks opens
with one brushwork of sand paint.
A mosaic of Tibetan skullcaps
on red, dry, stone mountains.
Flesh pink sunlight and blue casts
slice the temple floor,
while chlorophyll seedling sprouts.

Did we come with fire and dogs?
or were we planted here by milkweed?
The limbs grow under tiles, broken tiles
from marching cockroaches burning the earth,
with tapping steps of flame.

II.

Below the navel of men while,
riding on camel mammoths,
we pencil sketch gooseflesh
as we start bane in black.
Weights and measures of things,
named with gravel hearts
in degrees of blood desert sprawl.

By adding yardsticks of veins
then pewter, rod-like eyes of palm,
Are we capsules of water?
splitting the ledge of this world
by merging coils of genes
waiting for bread and wine?

III.

There are a million galaxies found
all drawn to Virgo Super Cluster.
Milky Way and Andromeda
collide like two ghosts,
spiral arms jetting and passing
through each other pulled by gravity.
Two black holes' cores circle then, merge.

At one end of the cotton turban
is sweat and salt, an imprint
of tones and paint. The other end
lies tucked away
in the fertile crescent of each man,
green or sand.

If a satellite can see the bottom of the sea,
then a picture hinges on everything else--
overtones of purple, white-hot lights
a few fragmented marks left etched
on the temple walls in ink and oil.

ON POETICS
Dg Nanouk Okpik

> ...*But finding no resting place, returned;*
> *then I sent forth a raven. The raven flew off,*
> *and seeing that the waters had decreased,*
> *[Cautiously] waded in the mud, but did not return....*
>
> The Epic of Gilgamesh

I.

When the mud dried black spruce culled
at the river's lapse, I slouched over to fill my mouth--
the ice pack gorge flowed over my fingers.
I cupped then drank. Right hand first, left followed.
Is this the way to the earth? I've stood still
but the sea and sky kept circling, circling
the midnight sun, I did not return.

II.

In the loft, I found one carved wing of yellow cedar
resting at the bottom of the netted cage.
Foul and cold air swept by me. Aaka called,
I dropped the wooden wing, fled down the ladder
to a black bird in a mask. A box of suet spilled.
I ran to the river to meet the ocean's edge.
I returned at dusk.

III.

Ellipse of the moon when the sun is the lowest--
harp, timpani, bass viol, flute,
wavelengths of woodwinds.
The nimbus darkened, a gingko fan leaf
measured candela carbon in the expanse,
Genesis at he dense blithe. The bell on the mountain
rang beyond the scape: echo, echo.

IV.

Blackfish parr, swimmer of freshwater--
urn of eggs pocketed in rocks.
Swimmer flow past in this moon
for you--brackish seasons leap.
So it is, you breathe quantum lux
and return, return, and return.

ODE TO ALBERT EINSTEIN
Dg Nanouk Okpik

What can I do A.E. now that the cold waste forms me?
A blue spotlight etching my bones ceramic.
I have digits strumming
a chord flat in ice.
I twitch in three layers of skin
my infected fibers sculpted
to hide the framework and music.
Now that unthinking spreads
so often like an electric switch;
on, turns the galaxy edgewise,
off, turns to black tunnels.
What fills nothing?
We hear nothing frozen in time clocks,
eight minutes is the length of the universe or is it?
A field of bodies, stark and naked,
necks turned sideways like life in quantum jump
star-like, half-life in Doppler effects,
big falling snow dust.
Will they make inkblots?
Will they think to tie the strings?
Extend motion and light a few black candles.
Will the gradual shift in planetary shadows
of bodies and clocks tick, tick?
Or will years, side by side, sundial and sundial,
rise in eight minutes and the beams strike and rest?

Color.

Hue of a life
I
never
knew.

Indian Child.

I
sing you
into being.

Silence.

I
raise you
into voice.

INDIAN UNIVERSITY: FOUR O'CLOCK
Sara Marie Ortiz

Indian children
I am knowing you.

Indian children
I am seeing you.

Indian children
I am speaking you.

Indian children
I am living you.

Indian children.

I am dying
you.

Indian children.

Frayed pieces of hot pink fabric
in the green of the freshly mowed
grass. Walking to you, I saw the
body there. Or half of one. The maggots
had already taken the other. The head.
Only a ringed, light pink, tail remained.
Only an end. No beginning. An end.
I should not have, but I did. Me
pulling gently at the tail, turning, and lifting
so I would see more clearly.

The little bodies, the squirming opaques
and movements of them. How quickly
they writhed and dropped from the hole
they had burrowed in its stiff flesh. How
quickly.

74

No sound.

My breathing, slow, and impressed. Poised
in the late afternoon swelter. The stones.
The head. They. It. Is gone. And I

brush the brown strands from eyes, hold
them away. And move on.

BLOOD VOICE
Sara Marie Ortiz

Swallow:

Genocidal bureaucracy
Frosted Flakes
and
Librium
with red, white,
and blue
sprinkles

breakfast
lunch
and dinner.

Wash it down

with half-breed blood
from a bone flask.

I got up
when they asked me to.

I got up
but instead of
going to the back

I went inside.

South Dakota
jail
and me.

Orange
county jumpsuit

warm
butterscotch pudding coating
the swell
of my pink tongue.

Black
bailiff
shakes his
black head

as he walks away.

After all he
was
only doing his job.

Genocidal Bureaucracy.

Spell it: A - B - C

No need for Wounded knee
anymore.

No need for Alcatraz
anymore.

No need for Leavenworth
anymore.

No need for Sioux Falls
anymore.

No need for "peaceful" re-conquest
anymore.

No need for Trail of Tears
anymore.

No need for Sand Creek
anymore.

No need for smallpox blankets
anymore.

America Online
FOX
Pepsi
Universal Studios
Chevrolet
Mattel

and

McDonald's...

get the job done quicker
get the job done quicker
get the job done quicker

Infect the young first...

the sick and the elderly will go next
the sick and the elderly will go next
the sick and the elderly

will go
next.

Infect the young first
get the job done quicker.

MAGIC
Sara Marie Ortiz

Coyote.
You owe.
The money was short.
This is the last time.
Coke isn't cheap, you know.
Just because you've gotten away
before, doesn't mean, you can hustle
me.

Coyote.
You owe.
The people are hungrier than ever.
You said the fire would save us.
You said we would survive, even
past the burning.
You said.

Coyote.
Your mom is a crackhead.
Creator is a whore.
Believe.
I said it.
Your mom is a whore.

Coyote.
Why did you trick us?
You said the corn.
You said the rain,
would come
again.

Born, from desert and of
ancient salt, you said
we would
be okay.

Coyote.
You owe.
2004.
Only 8 years left.

You owe.   79

41 MILLION
Sara Marie Ortiz

Urban Indian nightmare
come true.

Bleeding and frozen dream
on a stick.

Indian boy
said he'd come home a soldier.

Nobody wanted to see him wrapped
in an American flag.

No one wanted to believe the first dead
would be an Indian girl.

No one wanted to admit the first dead
would be an Indian girl.

Coming down to it
and now I see.

The flag is a blanket.
The red, white, and blue
is a tapestry

of dead Indian children.

And 41 million could never be enough.
And 41 million will never
be enough. Because

American women's wombs
are hungry.

American women's wombs
are hungrier than yours.

And.

The flag is a blanket
to wrap
the Indian boy in.

He said he'd come home
a soldier.

TOOL
Sara Marie Ortiz

Never
really knowing where the story will take me.
Just knowing, that the story
will take
me.

I am
a child. I
am a child.I
am an Indian
child.

Born to tell the stories.
Born to bleed
the stories.

You thought this
was about words. You
thought this was about. You
believed, there were no Indians left. You
believed.

Alice, never dreamed this.

I.
Spoke.
You.
Into.
Being.

You.
Died.
Anyway.

INDIAN LAW
Sara Marie Ortiz

Feeling the need
to list
all the things, I cannot list.
I cannot code. 100 proof. 2004.
Too harsh, for an Indian girl's
saccharin poesy.

He touched me.
She slapped me.
I stole from him.
I tore the tissue doll, limb from limb.
He left me.
I found his condoms under
the sink.

The white women were never alone.
There were lots of them.

Feeling the need.

I inhaled lots of smoke.
Blue and Green.

She
bought me my first bottle of wine
when I was 5.
He
taught me how to ride.

Alcatraz in 2001.
Wounded Knee in 2002.

The bailiff remembers my face
but not my name.
She taught me everything I know
about the code.

XIT.
Fame.
Indians
Famous Indians
make little Indian girls cry.

The studio is a cyphon.
The book is a prison.
The stage is a blanket.

I see you.
Tribal authority.
The bailiff remembers my face.

The Indian judge remembers my eyes.

Imagine, if I just started writing. Imagine, if a I just started writing about, Tupac, and Peltier and the rest of the serpents. Imagine. Imagine, that the letters weren't serpents and skeletons, dressed in black suits made of Dine` baby hair. Imagine.

The night is so loud. Sometimes, I wonder. Sometimes, I wonder if you can even hear it. Streetlight. Light in the desert. She did not imagine you there. No mind.

A truck in the distance. A dull roar. Carrying something away.

Carrying everything away.

Imagine, if I just started writing. And the world came alive, instead of

what it is doing now.

Imagine.

Imagine if, Geronimo were not dead, and you

were never alive.

Imagine.

I started telling the story, at 1:06. Morning. 1982.

But,

nobody could hear.

Brilliant
bleeding
neo-lithic
rays.

Clique`?

Heard the owl

DOPE FIEND

in your Indian eye's
ripping
today.

No one heard.
But I told the story,

anyway.

CHRISTMAS DAY: 2003
Sara Marie Ortiz

Fluid motion.
Ever pressing
forward   motion.

Light.
Like no other.
A brightly burning
ember.

No time to shine.
A Lakota boy
writes lullabies
by firefly light.

Another Indian mother,
is passing.

Another Indian baby,
perhaps a girl,
is coming
into the light

of an IHS delivery room.

Light.
Like no other.

Jingle, jingle.
Another Indian baby
has just come into

the light.

An ever pressing
forward   motion.

INDIAN OUTLAW POETRY
Sara Marie Ortiz

Bury my heart at Wounded Knee
feed my eyes to the ghetto
throw my Indian mouth
out to sea.

The Seventh Generation
miscontorted, distorted
as an infestation
but coming
as a
revelation that
annihilation afterall
is sometimes necessary.

Silver rings and
turquoise linings
come to me in dream
but leave with reality
on the underside
of your tongue.

Have you ever seen a real Indian?

Have you ever seen an Indian play the violin?
Have you ever seen an Indian
out of jail?
Have you ever seen an Indian recite a poem,
or bust a tight flow?
Have you ever seen an Indian pick
his nose?
Have you ever seen an Indian
cry?

Have you ever seen an Indian?

When will the crashing cease?
Oh when will the crashing
cease?
Oh when?

Burning shine
"this white-hot light that I'm under
must be the reason I look so sunburned"

History lesson taught
by a man the children know as Slim Shady.

Indian Sonnett written
by violin's illume.
Dying poet's breath
on my neck
and
I won't be done for some time
No, I won't be done
for some
time.

511 years and counting
511 years and counting
511 years and counting.

Did you think I had
forgotten?

Mercenary for crumbs
fragments and morsels
of light.
Ink blot looking
more and more
like an Indian man
leaving.

Have you ever seen a real Indian?

Have you ever seen an Indian make a pizza?
Have you ever seen an Indian weave a tapestry
out of trash?
Have you ever seen an Indian
raise their hand

in class?

Have you ever seen an Indian reading Hemingway at the beach?
Have you ever seen an Indian braid her hair?
Have you ever seen an Indian dye
his hair blonde?
Have you ever seen an Indian explain
American
history?

Have you ever seen an Indian?

And when your hair falls in your eyes
they think you have nothing left to lose
they think you have nothing left to lose
And when you hair falls in your eyes

they think
you have nothing left
to lose.

Red Star
violin native showcase
Ten Little Indians in
army fatigues
E&J in the car.
Lakota Millenium nation
deliver me into an
Indian tomorrow.

She said: "I feel so far away."
Tied her brown hair back
with the strings of my Indian heart.

Watch the Indian marionette show
pay
only 50 cents and a strand of your blonde wig.

I can't escape, but

in my eyes
you can still see
that I want to.

Have you ever seen a real Indian?

Have you ever seen an Indian
seeing
you?

Have you ever seen
an
Indian?

Bury my heart at Wounded Knee.
Feed my eyes to the ghetto
throw my Indian mouth out to sea.

*written: October 2003

MA'HEO'O
Deleana Otherbull

Out of my jaw, he made graceful pony beads
Slender, colored, needle grasped into leather
From this, he sinewed my life
Into thick strips of colored quills
And concho belts.

From my hair, he made dreams
Dark, elegant, twisted into metal cones
From the strands, he braided my thoughts
Into fine rows of horse mourning
And anguish.

From my teeth, he made spirit trails
Unlit, hidden, in mounds of crooked deaths
From the enamel, he cemented the tracks
Into firelight of lonely walks
And soul steals.

From my fingers, he stitched the world
Soft, beaten, uncurled to inhabited tongues
From the length, he made day
Into nights of squandered lives
And sentiment deaths

BOARDING AT SCHOOL
Deleana Otherbull

Curdles of beaded looms
slip through my fingers, like grains
of hot peppered time
undisturbed to the sound of old peyote drums
and purified naked sweats
of aged women and smelly men.
My hair was cut, short, into blue uniforms
and slaughtered buckskins.
          Note: *I didn't cry when they hit me.*
They taught me how to believe
to hold my pride between my sores
and let salty dreams
lazy stitch themselves into dirt rows
of moist mold.
          He does the same thing.
Inside his hands
he molds my hair into mourning
of treaties and allotments
folded between my ears and divided,
three times like a burnt flag.
He's a trickster, unseen in yellow satin
shawls
and orange flames of bright ribbons.
He glistens in burnt porcupine quills
and snowed in death songs.
I am his alien
to my own land, acres, of unexcused absences
and allotments I never owned.
Illustrations of fired revivals
ghost dances held by the dead,
he tells me to wear this shirt
and they will never see my pain.
And I do. And I hurt
but my beaded pride
will never surface.

HANDGAMES
Deleana Otherbull

My toe tips on the spine of your spirit, gently,
forcefully, like a man's braided coup stick.
Underneath bridges of battlefields and ghost riders
I come to you, with cut locks of my dignity.

Turn East towards the sacred hill, listen for death calling
and let the fluff of the bread take you alive.
Crisp with ancient leather blindfolds
instead, I will take you as my own.

Old women, stitched into lazy bingo cards
row themselves up into fancy ribbons
and lacy dust of my moonlight gaming.
Tell them- I didn't want to play anyways.

You: tell Ga`a that I danced like a man
with my high leggings and ribbon shawl:
I was the plume in the wave of her hair
and the willow stick between her soiled fingers.

Me: I am the cry of your victory song
salted with unborn infant amulets and butterfly kisses
bobbing to the stick of your Crow hop legs
untamed. I have become the anguish of your contests.

MY UNCLE
Deleana Otherbull

I catch the tears that you have shed
into thin buffalo stomach sacks
and horn carved bowls.
I string your tears onto sinew threads
lining thin rows of telepathic stories
and porcelain covered roach clips.
The pine quills shudder, as I scrape them
off of your wrinkled body
with wax and fancy needle work
made from years of decomposed fingernails.
Your bussel is my entrapment of lies
sneaking up on me over abandoned plains
and war painted ponies.
        Close your eyes, I tell you, *this is a dream.*
Full of make believe visions and obnoxious figures
I try to tell you what to think.
Through the belly of my amulet
I tie your tear drop necklace around death.
From traditional scalps, I cement my tracks
with enamel. Lonely walks and needled tongues
tack down your memory.
        Let us wrap ourselves in their blankets
        of diseases and parched lies
        for I no longer wish to live
        in their world.

95

NESTLING
Deleana Otherbull

We bend the smoke
across chokecherry twigs
and broken *ho`toa`o* bones.
We ask for blessings
tied cloth over mounding trees
ribbons laced with dreams.
> I never knew
> I was the plains
The curving of water
the looseness of the past
wrinkles into my features
of dew musk and
open stomachs
of commodity cans.
Where sunrises fall behind us
and the doors
close
in front of us.
> I never knew
> I was the plains.
Our tipis face to the east
blocking the rays
of withering terrains
that break us down
into burnt bushes
of storytelling and games.
We hold this passion in
when the passion
lets us go.
> I never knew
> I was the plains
In my car
67,000 miles with no history
we chase ancestors for fun
remember the time
over spine bumps of the land
and pieced memories?
We rolled the window down

I am not a poet
Filled with words to describe
Something I never knew.
I am the plains.

PAPERBACK
Deleana Otherbull

She was:
The crude sashay of emerald embers
glistening on the stitched clay walls
of no money and no water.
The potential cracks endure her life
artistically drawn into deformation
of lost dreams to desert winds.
Small compressions of yesterday's veins
hold up while she breaks down
into a mirage of rhythmical lights
dancing along the wash of movement.
Small tokens of candy line her pockets
like eager nimble nails, poking through
skirts of happiness, like unbroken horses
eating the last scenes
of evicted traditions and George Strait glories.

THE DIVIDE
Deleana Otherbull

Like the shallowness of your heart
I plunge deep
into the heeding *heme`konema`sehan`eeveno*
where the buffalo still graze
and old covered wagons
still blemish.
Slowly
falling from tattered pipes
and soiled waters
I grasp the hollow rock
which catches the mist of *mah`pe*
and still born barrels.
What is it like?
To almost taste dampness
from the creeks of Big Horn
or to dwindle off
and catch yourself
underneath the haunting Sarpie's hills?
They say that *he`eo`o*
can't touch the feathers
of an eagle
or drink straight
from the springs.
But as I lean down
I hear the water rushing through
and the feathers
falling for me.

heme`konema`sehan`eeveno means Crazy Head Springs
mah`pe means water
he`eo`o means woman

99

NIGHTMARES
Deleana Otherbull

Envious fists of skin
welting over with
morning rays
of pow-wow songs
that fix themselves
on the wall
like old history
and beat up
Fords.
My grandma makes pudding
the gathering kind
out of ripe chokecherries
truth of authentic
beadwork
and championships
I never won.
I try to eat
off the fire
of warm belly soup
and dried deer meat
with no intention of
crazy Indian ways
and drinks filled with
bruises.
My grandma is weak
humbled over by years
of shame
and torn into little pieces
of yucca
from the woman
who named me
Muc`sinut
"Red Woman in the Creek"
        a woman
drunk by reminiscence
and pulled together by fear.
A woman.
Wild onion roots
Gathered by fingers

made of
Medicine bundles
And burnt sage
where hunger is an option
mother
left for dead
on a prairie
filled with memoirs
where lives create passion
and death creates
survival.
I am not a woman
sworn in by clan
and nursed by a tribe
made from name and air
I am a child
born into calling and water
left for the Big Lodge People
and hungered over
by love.
Yes, I am a child.

CONFIRMATION
Penina Partsch

Her lips move slowly and deliberately.

In the corner, a man begins to dance, laying grotesque shadows on the wall. He moves to the rhythms of the bass guitar pulsating on my chest. To the left, a man; slicked hair, spits on the face of a young drunk girl as he charms her. To the right, a woman; too much eyeliner running down her cheeks from laughing, cackling so loud over each metal string slapped and struck, resonating deep into my head, vibrating my eyes. I look anywhere but straight ahead. I search each face in the tiny bar, each face of too many long nights, too many children, too many years, too many drinks. We all struggle the same, struggle for the same air,

But not her.

She just leans through the umbra and continues to move her lips so slowly, so deliberately. Yellow breath of an exhaled cigarette cloaks her eyes, but they dig into me, even deeper. Her mouth begins to press forward. Words float between glossy lips like sugar coated moths, gripping the back of my throat. I brush my hand through the smoke, searching for eyes to prove me wrong.

A light breaks between writhing bodies on the dance floor, caressing her face so gently, so intensely.

I watch as eyes and lips mesh.

I watch as she breathes venom.

I enter through the splintered doorway of my home. There are no lights. Eyes shut, I walk through the darkness, around the wooden dinner table, past the antique couch, through the smoking room and into my haven. I know these hallways, these rooms and these doorways. I turn on the light. Outside the wind hisses on, spitting leaves at my window. The house feels so hollow, so empty, but the floor boards creak. I close my door tight and prepare for sleep. The light turned off, I wait watching shadows laugh on the curtains. Louder and louder, the leaves scream and scratch. And then silence. I wait, eyes and lungs motionless. The floor boards, why aren't they creaking? The wind is not hissing. The leaves are not moving. I begin to rise, to see the calm and then – it roars again, bellowing of caution. The wind shakes the window throwing leaves and shadows unsparingly, violently. I can't hear that:

Sibilated hexes ripple the air stopping at my doorway

THE BOOK OF LIFE
Penina Partsch

Adrift amongst the evanescent
Shadows bow to grieve
Upon my uncertain shoulders
Come the

Copper

Clatter

Cascade

Time lies askew from its tracks
Laying black upon black
But revolution wails for exigency
Should the

Clairvoyant

Coins

Commove

Linear visionaries might beseech:
Come battle the cosmos
Atop your brittle ankles

WAVELENGTH
Hoka Skenandore

The speech of lungs cuts
one particle at a time.
Thin sheets of displaced molecules
pass silently through gyrating ions.
The double helix of the soul
is split into fractals.

These pieces settle
and eventually rot into a fleshy stew,
thick and slow.
After festering the amalgam bubbles
and pulses back towards an invisible center.
The singularity is reached.

CHECKERBOARD REFLECTIONS
Hoka Skenandore

It was a simple matter,
a mix of
chess, reggae, and gangsta rap.

Two moves ahead,
minus one back.
Greasy finger prints
on slick black grooves.

Pawn takes rook.
Queen takes pawn.
Slow steam cloud rises,
to be lost in cigarette smoke .

Queen moves left.
King finds quarter,
in a house of pawns.

Eyes dart about,
looking for keyholes
to shove bishops through.

Jamaican drums sound
the call of war.
A kings abode
becomes his tomb.

Time to make tea
and think about
last moves and tomorrow's note cards.

"T.T.T."
Hoka Skenandore

It's the best bartering tool
      common currency for the deprived
another simple acknowledgement of human existence.

Another unspoken code
      admonished as well as it is received into
the next ones shaking hands.

Best used when
      it sends its patrons into fits
like a maddened phoenix scraping at its own ashes.

It got grandfather
      his daughters and some of their descendents
and close to their death-beds they still embrace the spirals
      toward heaven.

It welcomed me with
      the same longing embrace
as those who seek its asylum, comfort and warmth,
      confined next to the lint in my pocket.

BODIES OF WATER
Douglas Two Bulls

Swim fast oh youth-- there are sharks in those American Lakes--
Inserting spermatozoa and feces-- Into Lake Michigan-- You are all
naked in a jellied salad-- Engaged in a harmonious orgy-- A passing
lead through a molar-just waiting for a serpent or a beast to
indulgently caress and bound you to its acidic belly--

A sinking possibility--

A corsage on cocoon garment--

A condolence from a contour drawing--

A missing month on a pin up girl calendar--

A personification played out on a woman's breast--

A luxurious concept--

Lubricated--

To be swallowed by a razor sharp jaw--

I scan this continent from an aquarium toilet bowl- playing war with
flounders and gold stamping illusions of boiling lobsters
on plates--

SOLITARY PEASANT
Douglas Two Bulls

As tinsel sways in ordinary nights
I build a monument with Abraham's copper
A night not stolen from burned embers

With clamped hands
I utter written tongues
For a moment to spend a secret
Upon a perfumed anemone

From a incubator
I recall marching Sky pilots
Into the wilderness
To be washed

Herod gave the head of Johannes Baptist
To reward a dancing girl

The talons of a sleeping industry
Hold siege upon my dire drones
Who fed nickel and lye
Into our queens larva chimes

To reset the genes of partial
Memory

ARMADA CATS
Douglas Two Bulls

Courtesans for armada cats

Nine lines in place of feline lives

Breeding violently in luminous corners

Inhaling from gaseous lungs of a physics package

Drinking from cascade beige to cocoa and sage

To fall asleep in my silhouette

BETA MINUS
Douglas Two Bulls

Aurora
Upon my claret bayonets
My sister a broken musket

The victor's heroes in a copper aviary
Peer out with paper birds
Enameled on an atlas and cutting board
Scotch taping images to their cranium
Gluing heads and tails onto obnubilated
Bullets

Coke can partisans
From a womb to a hill side tomb
My sister a broken carbine

PISS POT APRICOTS
Douglas Two Bulls

Piss pot apricots

In steaming pools of stew

Hesitating fermenting

Moistening silhouette fingers

With the stinger of a yellow jacket

Parading kitchen knives

In place of kid cotton crayon

Slaying fragments of bacteria

Under a constable's fingernail

Giving light upon:

Cotton sheets

A WALKING LAMENT
Douglas Two Bulls

From your door to mine
I cross puddles of water that hide cities
From Dresden to Vesuvius

From your door to mine
I dramatize an aerial view
Of fragmented trails

A maze to a chiseled artery

From your door to mine
Ions of Alexander protrude from hand
As I façade your paper skin barricade

SENTENCE
Orlando White

Look:

> paper screen

> blank;

> the color white,

> > a zero,

> > hollow light bulb,

> the O   not yet typed.

This means

> > no imagination
> > without
> > its *imagery*.

Letters   can appear

> > as bones

> *(Do not forget the image)*

> if you   write with   calcium.

Because a subject

can be half a skeleton,

the verb, the other half

and the skull,

> a period.

Head soaks in bleach.

Separate bone from skin.

Slide out the skull;

feels like Styrofoam.

Look.   It is a bulb.

Now,   twist it

into   a light socket.

When you

flick the switch

              a question mark

              should appear.

As for the white sheet

of skin?

Wring it out.

Hang over   the skeleton

              of a letter.   Let it dry.

LIGHT BULB EYE
Orlando White

Type.

Between   black

                 white

        letters   exist.

Words   _will_   soak

                        through.

Cursor   pauses

                before it

                flashes   darkness.

Don't be afraid

                to see   _the color_____._

Light

        bulb   inside

        socket   of skull.

Wipe

        the alphabet

        off   fingertips.

Except

the O   peel it   off.

Put it   on eye.

Zero blinks.

SLEEP AND LIGHT
Orlando White

The eye touches   chlorine surface.

   An O bends its knees

                moves across   paper.

     Its tail   a vital sign.

        The light bulb   begins to ripple.

     Ringing inside a circle.

Bleach   discoloring.

     White heat   under   skin.

       The eye begins to   vibrate.

     Eyelashes   tremble.

Letters fold   between   breath.

     In the center   of a zero   an echo.

Head   moves

across   floor.

        Ears   fall off.

        Pick   them up.

        Put   them back

        into   the face   of an eye.

Tongue on   pause.

    Shake

        blank paper:

        no   letters.

Use   white   ink

        to write   with,

then   read it

with   a lens   shaped

        like   a light   bulb.

Watch   a zero

roll under   darkness.

        Blink   a caesura.

FALLING ASLEEP
Orlando White

Light bulbs inside the sockets of my skull burn out.

The O begins to dissolve inside the center of a zero.

Clock in the shape of a skeleton cracks its knuckles

If you press hard enough, a letter can be made back to an ink stain.

Bones soften to Styrofoam and snap under the skin.

Everyone is a lower case i or j clothed in black suits and dresses.

The skeleton separates from out of a shadow that was bleached.

People are sentences and they hold either a period or a comma.

Silence can be the sound of scissors following a dotted line.

I use ink to draw myself and then cut myself out from paper.

Someone erased the alphabet, which later, appeared in my head.

When a light switch is flicked on; the skull smiles like a light bulb.

I look at myself in a white mirror, my face, still a blank piece of paper.

I eat a clock as if it were a watermelon, and spit out the numbers.

Silence can also be the sound of gravel being stepped on.

I took a class on dreaming and wrote my notes asleep.

I threw my skin into the laundry bin after I undressed.

I made love to a skeleton under the covers of a zero.

Have you ever seen a blank sheet of paper burn out like a light bulb?

I was born and my head appeared like a speck at the end of sentence.

Erase   numbers

        from   clock,

shake   the O

        until   white   appears.

    Put it on face.

You   are not

blank

or   a skull,

but   a black suit

shaped

            like   a letter

whose   face,

        inside   the dot

of   an   i,

                        is   a zero.

The shadow puts   a circle

over his face.

                He does not know

his skull

                has the color of Styrofoam.

He knew   under his skin

        a skeleton

rubs like   paper against ink.

He found an egg   and opened it.

        Shaped   like the letter   O

it was   empty.

He liked that

                because a zero

cannot blink

                but can be   filled in.

LOWER CASE i AND j
Orlando White

Man

with   one leg,

    no arms,

    wears   black suit,

        white neck tie.

Woman

in   black dress,

    white scarf,

        no arms too.

On   white

    sheet

both   attend

    paper funeral

    then ink   tears.

But   you

can   put

     a hyphen

       between   them:

    look

they are   holding hands.

Soak eyes

    in white.

Erase   ears.

Listen   with O's

    on head.

Put nose

    in bleach.

Sniff   discoloring.

    Peel zero

    from   a page

    eat it.

Does it taste like Styrofoam or a tooth?

Rub   blank paper

    inside   sockets   of skull.

Remove   hands

    place

    in envelopes.

    Drop   them

    into   the mailbox

    of   a circle.

A punctuation

    mark

soaks   through

    white.

Press   ear

    onto   blank

      paper:

    sound   of space

    between   letters.

Dot

    with out   i,

period

    with out   sentence.

Use   ink

    shaped

      like   a hyphen

    instead of

      zero.

Listen.

It will appear

on the page.

FROM SKIN TO BONE
Orlando White

Vital sign of a patient,

check marks in boxes,

life, death. Still eyes

behind paper, film.

I touch my Grandfather's

hand. Breath trembles:

silent movie. Bleach

diluted with darkness.

Eraser smudges,

it does not erase. I read

his eulogy in a crossword

puzzle. Watch black,

white television, static

shapes, the *yin* and *yang*.

A skeleton holds

the last sheet of paper

and I wait for the cursor

to stop. Today's newspaper

said, "Tomorrow's Suffering,

Yesterday's Grief." On my

page I see a checkerboard

of fill in the blanks.

I use punctuation marks

and letters to play it.

But, I shiver at the clock

ticking over my shoulder

shaping itself   into a skull.

Skeletons   shaped

like two

    letters.

She   holds   a comma.

He   holds   a black dot.

Both   do not   know

    each other

until   they are

laid   next to

    one another.

Underneath

    the paper

    both have   faces.

If you   peek

    under

    the page

you   will   see

two   lovers

         shaped

         like i and j

kissing   with   a hyphen.

FICTION

ARSON
Britta Andersson

Maria is the daughter of the son of a father planted beneath the corn in the mountain village of Nebaj. Her grandfather, now part of the harvest she grinds to mix with water, is tapped between her hands a pleasant song. Behind the woman pressing oranges into juice, a church is draped with caution-tape the color of fruit. In its cellar, amidst the deadened air; lay bones twice scorched by flames.

The shade of a barbed wire fence falls upon a little girl' face; it casts a black rift of pointy stars. Her blouse, an extension of her face, shines of hand-stitched rainbows, purple flowers and, golden blooms. Intent, she observes her mother hard at work. Her worn hands moving over and back, under and over, stitch sunlight on cotton. Mid-day, as the girl gathers wood for her mother's fire, the sun's stitched rays bend in her skirt. They glimmer orange paths that extend to the morning star.

Barefoot men and boys play soccer on a cleared dirt plane. "Goal!" shouts the smallest boy, so passion full, that his voice shakes rain from clouds. On the sideline, an old man is content to listen; child sounds regenerate freshness in air. He remembers the day the air went stagnant, when trees made no noise, and vultures swarmed black circles in the pale blue sky.

Twenty-four years prior, the army forced the coastal men of Guatemala, to strip the men of this village to their skin; parade them bloody in front of children and wives. Their mothers' voices muted with the wind. The mountain men, portrayed as "rebels", were doused with gasoline and set ablaze. They were buried in a shallow grave, littered with mismatched shoes and a plastic doll. Corn emerged years later.

Vendors sell rice milk, hot chocolate, tamales and atole, black salt, mangos and corn; the street market chimes with trinkets and bootleg CD's. A child asleep on her mother's back is wrapped in a collage of red feathers and green wind, yellow corn and white-shaped stars; a woven tribute to what lingers beneath the tall grass.

Quetzals hid in the forest the dusk that vultures arrived. Their lustrous feathers shaking loose, their chests heaving to take in air; shocked by the shouts of, "FIRE...!" and bayonets pierced through ribs.

It was before the bells chimed twice in the dim lit air that the match was cast inside; the evidence against a sergeant drenched with flames. The altar lingers of burnt commandments and marrow. That February dawn, exhumed bodies recorded by the church turned black, escaping stained glass windows as ashen air. Death threats ring the priest nightly.

Maria pats tortillas from palm to palm; wind wrestles her long dark hair; the smoke chimes lightly against her skin.

There are eleven strangers in a classroom, split open like cracks on skin. Cracks, meant to reveal some kind of ordinary fate, yet resonate like a white sleep in a room that exudes no darkness.

We embrace a common question, born from the need to discover the intent of our pulse: Who defines the sound that escapes our hands?

As young writers, we are still foreigners to ourselves. I search for myself: a shadow in a fluorescent-lit room, hidden beneath cold skin, reflecting cheap energy. We read black songs cast on a page.

Girl 2 enchants the white walls with a collage of whispers from her younger years. They coil against each other like snowflakes on an ocean's edge.

Boy 3 parallels a moment with the scent of sperm trapped on vinyl.

Is this what happens when we look inside ourselves? Something like a whisper appears on a screen, attempting to shout what survives in the back of a heart. I wonder what it translates in a language not yet conceived? Does it sound like melting?

Boy 5 sips on ice from a Styrofoam cup, and swallows at the rate of Girl 7's comment.

Girl 7 craves breadth, although the length of Boy 4's experience was "powerful", his range poignant, and the heat of his encounter "rich".

Boy 3 intrigues me. I always read his words first, imagine his voice escaping print as a cloud of smoke in water.

Girl 1 speaks with a hushed breath that forces her eyes towards the floor. I guess this is why she never looks at me and avoids the use of color.

It occurs to me that an image based mind retrieves a silent narration based on sight. Dialogue creates invisible people, retrieving voice from memory or the company of ghosts.

I cannot escape the sound that escapes my veins. Only letters fall silent. The words are read aloud.

I think we listen to what is mute on paper; discover ourselves attempting to scream between white lines.

KARMA AMONGST THIEVES
I.M. Antar

~Adrian, Rapid City, South Dakota, Wednesday the 16th Noon~

Adrian rolls off of the couch at about two p.m. most days of the week. For the past two months he has been staying at Jeremiah's apartment. Some nights Adrian comes home with a thirty-pack of cheap beer, puts it all in the fridge then orders whoever is home to drink with him. Jeremiah and their roommate Myles are generally happy to oblige. Other nights Adrian throws parties with kegs, and ounces of pot upstairs, and downstairs are the sex and "real drug" rooms. At least that's what Adrian calls them. He only trusts 'Miah with his money because the man has to come home eventually. Now this isn't because 'Miah is a bad guy or anything. He wouldn't ever try to rip Adrian off, but it is safe to say that in his own special way Jeremiah is an idiot.

Adrian hasn't had any business, other than selling pot for nearly a week. He is stuck at the apartment waiting for A friend of his from Arizona to call. Guy named Mike. Mike used to live in Rapid and has become one of Adrian's new hook-ups ever since he moved south. Once Adrian said that he felt uneasy about going through someone younger than him, then Myles reminded Adrian, that when they were 18, or 19 years old that they used to have a 14 year old kid get them their weed.

Mike started setting up deals for Adrian to pick up coke in Denver almost six months ago now and things are going smoothly. Twice a month Adrian Makes the four hour drive, picks up About ten 8-balls, and drops off the money with a middle man named Brice. With his hook-up's he is spending about a thousand on drugs each trip. Bringing it up north he can triple his money, but he prefers to get rid of it fast, so he generally makes between two and three grand. This month on the second pick up things are a bit different. Adrian can't leave town this week. Normally he picks up a new batch on the 1st and 16th of every month. Adrian Has family in Rapid. His parents and little brother live on Ellsworth Air Force Base. Adrian's little brother's birthday is on the 14th of November, and Adrian's parents expect him to stick around and help out. So Adrian sent Jeremiah on this run. He is only letting Miah do this for him because Miah has met Brice before. Adrian only trusts Miah to come back because, the apartment is in his name, even though Jeremiah owes Adrian two months rent.

It is almost noon, and Adrian begins to shift his weight. His

139

tightly balled fist comes up to rub the sand from his eyes with the shifting of his weight Adrian is thrown off balance, and hits the floor with a thud. He pulls the brown, tattered blanket over his eyes and tries to fall back asleep.

The ruckus made by Adrian, wakes up Myles who is asleep in a great big blue lazy-boy recliner. Myles tries to go back to sleep, pulling the American flag he is using for a blanket tightly under his chin.

"Hey, I know you're awake fucker! You got a cigg?" Myles asks.

"No." Adrian sits up, "Go buy some."

"I will, I will...So when is Miah supposed to be getting back from Denver?"

"I don't know man, hopefully tonight. The fucker was supposed to be back yesterday with my shit but I ain't heard a damn thing from him. But he has to come back sometime. He has everything here, his guitar, his stereo, his TV. He'll be back. I know he will. And he'll have my shit."

Adrian throws his blankets to the floor as he stands and crosses to the kitchen window. He looks outside above the dumpster, past the back parking lot's gravel top, between the trees perched above the muddy ally to see the beginning of a lazy overcast day.

Adrian turns from the window and grabs a black *Slipknot* shirt from the back of one of the brown, imitation leather kitchen chairs. He puts the small shirt on and it stretches tightly over his shoulders. Then he bends down to pick a black hooded sweatshirt up off of the tan linoleum floor, he pulls it over his head.

"Well At least I got my sack of buds to keep me company." Adrian says as he begins to pack a bowl into the "arctic blue" glass pipe that sits on the center of the living room table.

~Adrian, Rapid City, Wednesday the 16th 10:00pm~

Adrian lies on the couch, the only light in the room is the blue flicker of a TV screen. The sound of Lisa Simpson playing her saxophone is shattered by the squeal of a 1986 Volkswagen Jetta with a loose belt.

Adrian jumps to attention. "Myles, get me my bitch stick!"

Myles opens the closet door to the left of his tattered lazy-boy and pulls out a straight billy-club the kind that cops used in the 40's. It's dark green and has "bitch-stick" written across it with a red permanent marker. He throws it to Adrian who stands by the door. He puts one end of the club on the floor and leans on the other end with his right hand, like cartoons of big burly baseball players leaning on their bats.

Jeremiah walks in with his head down and a cloud of tears blocking his vision. He has his hood up and sees only Adrian's feet, just to his right as he enters. He lifts his head just enough to see the staircase descending in front of him.

"Where's my shit man?" Adrian demands with fury shaking his vocal chords.

"I don't got it."

Jeremiah sees an explosion of white with sharp points encroaching to the edge of his vision as Adrian brings the club down on the back of his neck. He hits the floor with a thud and is still semi-conscious. Adrian grabs him by the back of the hood and throws him down a flight of stairs that descends a few feet from the entryway.

"What are you gonna do with him?" Myles asks.

"You're going to get me that fat roll of green duct tape from on top of the fridge then your going to hold him steady while I tie his ass down."

~Adrian, Rapid City, Wednesday the 16th 10:15pm~

"Alright, throw the bucket of water on him." Adrian commands Myles right after the last strip of duct tape is wrapped around Jeremiah's ankle.

Miah wakes up. "Man, I knew you were going to be pissed but fuck man!...Just give me a minute to explain myself." He says this with the kind of grog that one would expect of a bear being woken during hibernation.

"Shit, mutha-fucker the only reason you're even alive right now is that fact that I want a fucking explanation. Where the fuck is my cocaine?"

"Well I got to Denver and I went to talk to Brice, your man out there...."

Before he can get any further into his story he is interrupted by Adrian's impatience. "Fuck this. I was supposed to be meeting people right now who want their shit. Myles, you listen to his bullshit for a while I gotta take care of people. I still got some greens that I shoulda been getting rid of instead of waiting for this punk ass mutha fucker." Adrian pauses and walks to the dark corner of the room with a big black floor safe, he dials in the combination and pulls out a cold silver Glock 9mm. "Myles, if you don't believe his ass, shoot his knees out. If you do believe him, get him some food. I'm out" With that Adrian is up the stairs and out the door. Myles turns to Jeremiah ready to listen.

"I think you better make this good man."

"Alright Myles this is how shit went down, And remember fucking Adrian said I could stay a couple of days and party... "

~Jeremiah, Denver, Colorado, Saturday the 12th mid-evening~

I got the shit and everything was all good. I saved a few bucks so I could get a room and party out a couple of days. I got my room and left the briefcase on my bed, it was safe there, shit I was the only one with a key to the room. I chose my hotel carefully. Actually I chose it based on the fact that there was a topless bar downstairs. The Volkswagen was running just well enough for me to make it to Denver and back. I mean it's only about a four-hour drive.

Anyway I decided I wanted to go down to the bar and see if I could get myself a few beers. I was a bit disappointed because it was something of a sports bar. But I decided not to worry about it, the place had a really good porter on tap, so I had about three or four of those before I thought to ask the bartender when the next dancer would be on.

He told me, "In about half an hour."

I looked at my watch: 5:30

When I looked up from my watch I noticed this mad, fine honey a few pool tables away, eyeing me down. So I walked over to her.

"Whadaya drinking?" I asked sounding as suave as I possibly could.

"Whadaya buy me?" She says without a hint of a joke in her mouth.

She smiled and grabbed my hand to take me to the bar. We put down Long-Island Ice Teas for the better part of an hour. She told me her name was Lauren, and that she was on her way to Tucson.

She said she wanted drugs. I told her I had pot, and she nearly dragged me out of the bar, asking for a place to smoke.

Now I'm man enough to admit when I'm buzzed up, but this girl looked damn good. I thought it would be a good idea to take her back to the room. I knew she would be curious about the briefcase though. I was a bit too buzzed to worry enough though. I didn't tell her to wait or anything I just opened the door and in we went. She saw the case right away.

"What's in there?" she asked curiously.

"Nothing, baby." Of all the things she could notice, she had to pick the fucking case.

I figured it was no big deal. Adrian told me if I was in any trouble, like if I needed to ditch the case I could drop it off at the gray-hound station, locker number 420. Adrian said he always had it on reserve.

I told Lauren I would be right back. But she said she didn't want to wait in that hotel room alone.

"Shit woman! You got free cable, a mini-bar, drink what you want I'll pay for it." But of course she wasn't having it. I thought it wouldn't hurt for her to ride with me to the bus station. We got there and I told the bitch to wait in the car. She said she would. I went to the ticket-station desk, and waited in line for about ten minuets. The station wasn't too busy this time of night. It seemed as if there were only people waiting either in line or sitting on a bench with their hats pulled over their eyes. I got to the desk, got the key I needed. I squeezed the green, plastic, diamond shaped key chain in my hand.

When I got to the locker and opened it up I noticed that it wasn't empty. Adrian didn't warn me about that. There was a brown paper sack inside. I looked and there was some cash, and some coke. I left his money, man, I swear I did. I just dropped off the briefcase, and I... well fuck man I took an eight-ball. I figured he couldn't mind. He wouldn't, and if he did I'd pay him back. So I dropped off the case and just left man.

Lauren was still in the car when I got back. We went back to the room and snorted. And snorted. And fucked. She said that she wanted it rough. She told me to bite her pull her hair smack her ass. I was fucking loving it man. She scratched me with these porn-star finger nails that she had. I didn't think anything of it man. We were just having a good time right. I had a few needles, and I personally wanted to blast some of the coke. I offered her some and she almost started to freak out on me. I told her to calm the fuck down and she did. She said something about never doing needles, she used to watch her mom shoot heroin. Eventually we were out of powder and out of liquor too. The only thing that I could do was pass out. I was probably out for 18 hours or some shit man.

~Denver, Sunday the 13th 8:00pm~

I woke up to the sound of the hotel room door being kicked in. There were five cops guns drawn, all screaming at me. Yelling; "Get up you lazy son of a bitch!!!" and other things that I couldn't really understand. I sat up and said, "What the fuck is..."

Before I could even finish what I was going to say a cop came up from behind me and threw me off the bed and to the ground. Fucker started reading me my rights. All I could do was scream, "What did I do?" as they carted me away wearing nothing but a pair of boxers and the blanket they had thrown over me.

When I was being booked I finally found out that I was accused of the rape of a Ms. Lauren Baker. I didn't even know her last name until I was getting fingerprinted for her rape. I told the police what happened from my side of things. They said I was going to have to wait.

~Denver Jail, Monday the 14th 8:00am~

On Monday I woke up in my holding cell alone. They left me in a holding cell all fucking weekend. Laying on my jail mat I noticed all the writing underneath the benches of the holding cell. Mostly tags. I touched them and realized that many of them were done with toothpaste. It took me a minute to realize that I was finally alone in this cell and I could freely take a shit. Sometime after I woke up and left my mark on the cold metal toilet a guard dropped off my breakfast and told me to get ready to move to population.

I ate my food and pushed the tray to the large steel door. A few hours later, I'm not sure how many, a different guard came by to ask me if I had an attorney or had spoken to a public defender. When I told him no to both of these he made a check with a red pen on some piece of paper that I couldn't quite read. In about fifteen minutes this guard and another came to my cell, they put me in shackles and led me to a meeting room where I saw my public defender for the first time.

"Did you rape the girl?" He asked without looking up from a file that lay on the metal table before him.

"Hell no! Everything we did was consensual. What the hell did that bitch say."

"Well she said you raped her. And with the bruises on her neck and shoulders and the vaginal tears it's easy to believe that she was raped. They found skin under her finger nails, is that yours?" He still hasn't looked up from the file.

"Yeah it's probably mine. She told me to bite her, the bruises on her neck and shoulders are from my teeth. She said she wanted it rough. So I gave it to her. She started scratching me and I gave it to her harder."

"Son, I got a girl that looks beat up saying she was raped in the hotel room that you were found in, a hotel room that was rented out to you. You can see how this doesn't look good, can't you?" At the very end of his question he finally looks up at me.

"Listen I didn't rape her, I didn't force her to do anything. I suppose it does look bad, but I di....." The lawyer interrupts me.

"Well that's ok son. You should already be out of here. Turns out that Ms. Baker skipped town. You can't think of any reason that she would..." I thought lawyers were all business but not this one. He just dismissed whatever idea was in his head. "No, never mind."

I didn't care what he was thinking, if he got too nosey he might find out about the coke and shit so I didn't push him on, instead I just asked, "So when the fuck do I get out of here?"

"I've already started your paperwork. When I leave the guards will come by take you back to the holding cell, and in a few hours you should be out."

~Monday the 14th 10:00pm~

So the release went well. My lawyer had informed me that the hotel I was at was holding my property in storage. He also told me that I could find my car in the parking lot of the hotel. He told me how to get to the hotel from the jail, said it would be easiest if I were to just take the bus. When I saw the light of day again the sun was pushing it's way through a crack in the clouds. I decided that I would rather walk back to the hotel. During my walk I wondered if Lauren knew about the coke.

~Monday the 14th 3:15pm~

It was a long walk through downtown Denver. I only wished that I had my money on me, because three different people offered to sell me some buds. It would have been greatly appreciated, but as I said I had no cash with me. My wallet was in my pants back at the room. The jail gave me a pair of prison issue pants, and a t-shirt. I'm sure that I looked like an escapee, but I didn't care.

I got back to the hotel and they gave me my property. I asked the women at the front desk if they had a public restroom I could use to change cloths. They did. When I reached into the pocket of my jeans, after getting them on, I found only my car keys, I couldn't find the diamond shaped key-chain for the locker anywhere. I went to the gray-hound station. I thought to myself that they must have duplicates.

When I got to the bus station, the women behind the ticket-desk recognized me right away. It was the same women that I had got the locker key from on Thursday night (or Friday morning, it was after midnight). She stopped me before I went towards the locker.

"Excuse me, Mr. Vincent?..." She paused and waited for me to turn around.

"Yes."

"Sir this note was left for you by your wife, she already came by and picked up your property out of the locker." She said this in a matter of fact manner, as if she knew me for years, and believed that that bitch Lauren really was my wife.

"I don't have a wife, I met that chick a few nights ago at some bar!" I couldn't believe that she had let some strange lady, that she had never seen me with before open that locker, with out even finding out if she was really my wife or not, But I guess that bitch had the key so... nothing I could do about it.

At this point the pulsating reflection of flourescent light bouncing off of the linoleum bus station floor began to pound my eyes. I got dizzy knowing that explaining this to Adrian wasn't going to be easy. I didn't know what to do. As I turned around I opened the note.

"Miah, last night was nice, but you should have stuck with pot.

I do crazy things on coke. Oh, well. You'll live, right? Well I'm not sticking around... Ciao Lauren"

I threw the note away on the way back to the car. For a moment or two I thought I could hunt this girl down, but then I remembered that even the cops couldn't find her.

~Lauren, Middle-of-nowhere, Arizona, Wednesday the 16th noon~

Lauren pulls into a small gas station in the middle of the Arizona desert. Her silver '95 Toyota Celica panting, wishing for fuel. She cannot tell if the gas station has any attendants working. She leaves her car parked by the gas tanks and starts to walk to the front door. She is momentarily startled by the whine of a rusted *Phillips 66* sign swaying in the wind. She looks as if she has been awake 30 hours or more. The bags under her eyes are reaching a shade of purple she hasn't yet taken notice of. The wind stirs dust into a whirling vortex as she turns back to the front door she sees a twenty-something, man walk out the door with filthy clothes on.

"Gas tanks are empty lady. Expecting another truck on Friday though."

As she looks him over she cannot help but think that he would clean up very nicely, be a good looking guy.

"We have a few rooms here though, could put you up till then. I'd even give you a discount seeing as how you don't have much of a choice. No gas stations for over a hundred miles."

Lauren pauses as if she is thinking it over. For a moment she surprises herself by thinking that the car would be a fine place to sleep. Quickly, she dismisses this idea.

"I'll take a room till then, how much a night?"

"Charge you twenty dollars ma'am. That comes with a meal too, if your interested." As he says the word "meal" his left arm twitches slightly. Lauren takes note of this and asks;

"Are you the cook?"

"Yes ma'am."

"And what will you be serving tonight?" She asks acknowledging the growl coming from the pit of her stomach.

"Got me some steaks thawed out, like I said ma'am meals free of charge, I can come by your room about seven o'clock with it if you say so."

"Yes, I think that would be alright."

"Well then miss, come right this way into my office. He turns and grabs the door by its tarnished brass handle, it may be the only chunk of metal on the entire store front that isn't rusted. He opens the door and ushers Lauren in.

Lauren walks in to see what appears to be the average convince store. This is how the average convenience store might look minutes after a robbery by a group of teen-age boys. One shelf towards the back near the refrigerators is overturned, and most of the shelves are barren, the floor under each shelf is littered with unopened bags of chips and candy bars among other items.

"What happened in here?"

"There was a big lizard got in here last night. I was chasing it all over the store. Musta taken me 'bout two hours before I got it out of here. Now that it's startin to get really cold at night, critters like that are lookin for a warm place to sleep." As the man says this he walks behind the counter and pulls out a large dusty tome. He drops it on the counter and the thud is accented by dust exploding into the air.

"As you can tell we haven't had a guest here in quite some time." He flips the book open about five pages in. The last entry is dated 1999.

"Well ma'am if I could just get your Jane Hancock here..." He says as he turns the book to her.

She pulls a regular black ink, ball-point, bic from her purse and signs the guest register.

"Well alright, here's your key." He says as pulls a key from under the desk.

"You see that building just across the main road there, this key will open the only door on the back side of that building. There is a phone in your room, if you dial zero you'll get me, Mike."

"Well Mike I thank you for your hospitality, but I think I am just going to take a nap as soon as I get into the room. I'm a light sleeper just knock a few times when that steak is ready."

"Cupcake, it's no problem. Enjoy your nap though. I gotta make a phone call."

Mike picks up the receiver of an old black rotary dial phone, as Lauren walks out the door, she hears him say; "Hey, Adrian...oh well.... so did your man come through...."

As Lauren walks into the dimly lit room (the thick tan Curtains

148

block out most of the sun) she drops a few small bags on the ground near a small round table with three wooden chairs crowded around it. Lauren flips a light on and can quickly tell that there is a lot of space that she has to herself. Like a large efficiency apartment with one room a king size bed and a TV/ VCR combination on one side of the room, a small kitchen, and a separate bath room. She splashed the contents of her smallest purse unto the table, she fished out of the pile, a small mirror, and a folded piece of paper.

Lauren Turns on the TV, takes off her jacket, and switches around until she finds a cartoon. Ren and Stimpy are on. She carefully unfolds the paper and slowly pours the white powder unto the mirror. After she cuts four lines, in quick succession she snorts, one, then two. Then she switches nostrils, three, then four. As soon as she is done with the fourth line she lifts her head quickly pointing her nose to the ceiling. She gets up and walks over to the bed; she plops down shaking. Then she curls herself into the fetal position prepared for an introspective nightmare.

Lauren is on the floor in front of the television playing with a rag doll. She feels the space around her now... In the returning. She can see around her as if through a panoramic of fish eye lens. Behind her in the kitchen her mom is sitting down on a torn slightly padded gray vinyl chair. Her mother's hair is a mass of dark brown tangled layers frizzed from root to tip. Lauren can see her mothe'rs scars. The tears in her arms. She sees a man that she doesn't remember tying a truncate around her mothe'rs naked pale arm, just below the shoulder. From here Lauren can see the needle, can feel it ripping the tender skin on the soft side of her elbow joint. She watches the plunger pull back and draw a cloud of red into the already milky liquid in the syringe. Lauren can feel the sting. She hears the tear.....

A sudden pounding Awakens Lauren. She looks at the room's clock: 7:10pm

"Come on cupcake! I got your dinner for you!" And the Pounding again.

"Just a moment." Lauren says, her voice cracking on the syllable "mo-"

"Aw, come on honey hurry up. These plates are kinda hot!" Mike shouts.

Lauren pulls herself from the bed and while talking to mike is putting the evidence of her habit away. She walks to the door, adjusts her hair and opens the door with a smile.

"Hi! Wow that food looks excellent." Lauren says barley looking at it.

Mike invites himself in, "If it's alright with you I'd like to join...hey cartoon network, I can see you're a women of taste...oh yeah... so can I join you for dinner?"

"I'm not really that hungry right this moment I was, still asleep. But you're welcome to stay and watch cartoons with me...for a little while."

Mike remains seated at the same spot were Lauren was cutting her lines. And begins to devour his food, As if in a hurry. While Mike is eating Lauren puts her plate in the fridge, and lays back down on the bed to watch cartoons, It's the Acme hour and it seems as if one of the coyotes plans isn't going to blow up in his face. Her mind wanders as she lay there. Before she knows it mike is sitting by her legs laughing as the coyote falls to the ground while the roadrunner stands in mid-air.

Lauren enjoys the company but is starting to get a bit uncomfortable. Just as she is thinking this, Mike turns to her and says;

"Not that I want any trouble brought to you but I'm glad you got stuck out here."

"Whaddya mean?" Lauren asks curiously.

"While...ok I guess you look like the type...I just, I get so sick of blasting alone." As Mike says this he reaches into his left cargo pocket, and pulls out a small pouch, it looks as if it could be a cell phone case.

"I'm just out here alone all the time. And most of the time I don't mind that it's just when I really wanna get lit."

He unzips the pouch and pulls out a black balloon.

"And I have had this shit, it's close to pure...I've had it for weeks now."

He pulls out a custom syringe and a fresh needle. Then reaches into his pocket and pulls out a spoon.

"No, no, no. That's ok cowboy why don't you do it without me. I'll watch it'll be fine." Lauren interrupts knowing what he wants.

"Come on! Baby please. I'll just give you a little blast. Just so that I don't have to blast alone." He says half-begging but very insistent.

"I said No!" Lauren stands to walk over to the table, but Mike grabs her arm.

"What did you say to me?" Mike stands and grabs her by her shoulders. "You, stupid Bitch!" Mike backhands Lauren and sends her to the floor. "You don't tell me No! What are you better than me? Can't have a little H in your system." He kicks her while she is on the floor.

Lauren has a trickle of blood running from her nose. She manages to stand only to be hit across the face again and knocked back to the floor. Dizzy her world begins to turn. She sees Mike taking off his belt. He hog ties Lauren as he is tightening the belt she begins to drift out of consciousness. She feels the poke of a needle in her ankle. The sting is like a lighthouse calling her to a place that she has always feared.

~Postlude, Thursday the 17th, 9:00 am~

Mike Decides he doesn't need to deal with a hysterical bitch coming down in a few hours, so he decides he needs to do something with the bitch and forget about her. He goes into the back of the convince store and gets a red gas-can, it is a full ten gallon container. It is quite heavy. Mike lifts it easily and takes it to Lauren's car, he empties the entire container into her gas tank and finds a piece of wood a little less then a yard long. He blindfolds Lauren finds her keys, and her unusually large stash of cash and drugs. Then he proceeds to drag her out to her car. Then he pops her trunk and stuffs her inside. She is still in a half-coma.

He drives car to an old dirt road in a fairly flat and completely empty area. He shifts into neutral and gets out, he props the gas pedal down with a stick and pops the car into gear.

Satisfied he turns and begins to head back to the gas station.

Once he gets inside he picks up the phone and dials a long string of numbers.

"Hey Adrian....Yo, It's Mike....How you doing man?....I know, I know....But I think I might be able to help you out....something happened....some bitch it....it was a mistake..... yeah...yeah....anyway...I ran into a bunch of extra powder for you...Yeah it was the bitch's.... Ummm....actually yeah, her name was Lauren..."

FOX 4-9ER
Jody Barnes

Burns leaned against the sandbags tensely. His eyes squinted up at the rooftops and the sharp corners of the alleyways. He asked himself how he ever ended up in such a shit-hole. He had signed on for a 4-year stint with the Marines, never imagining he'd ever see any action. Then some uppity dink country pissed off some other dink country. Unfortunately for the uppity dinks, the other country had oil, or gold, or bananas or some shit that Burns couldn't remember. So America came charging in like a big, bullying brother. And now here he sat in a foxhole with Cpl. Kelly. Burns glanced over his shoulder at Kelly who was napping at the bottom of the hole. Kelly was the perfect example of how unfair the universe could be. Kelly was built like a Greek God; he was blonde haired and blue eyed with a fine straight nose. He had no scars on his face, whereas Burns was skinny-some might say scrawny and he was still fighting a case of teenage acne. He could have used orthodontia as a child. As his washed out eyes studied Kelly's profile he frowned and pulled his gun a little closer, like a child finding comfort in a stuffed animal. Kelly's radio squawked, he was awake immediately.

"This is Fox 4-9er. Over." Kelly spoke into his mic.

"Bring it on in boys. Over."

"Over and out." Kelly clicked off. "Well, Burnsy...ready to call it a day?"

He grabbed his gun and rifle and vaulted out of the hole. By the time Burns climbed out of the hole, Kelly's easy stride had taken him a few yards in front of his companion. As Kelly heard Burns struggling to get out of the foxhole, he laughed to himself. 'Burnsy,' he thought, 'good shit, but how he ever made Marines I'll never know.'

Burns hurried to catch up. He didn't like to be left behind. As he stepped forward something jerked his right foot hard. In a moment of amazing clarity he realized he'd stepped on his boot string, he was falling on his face and there was nothing he could do about it. A thought shot through his mind: 'Shit! Of all people to fall in front of, it had to be Kelly.' He hit so hard his head roared and his right ear rang. Up ahead there was a red explosion in Kelly's brain and Kelly was no more. He collapsed forward. His hand, not knowing it was dead, twitched once and was still.

Burns lay there with a mouth full of grit waiting for the laughter, the heckle, but it never came. He raised his head. All he could see were the soles of Kelly's boots.

152

"Kelly?" Burns called out. No response. Burns scrambled up and hurried over to where Kelly lay face down. Burns noticed the small, bloody hole in the back of the blonde head. "It's not bad! It's not that bad!" Burns reached down and rolled Kelly over. His face was gone. What remained was a bloody, horrible mess. Burns screamed and dropped the body back.

"Oh, God!" He started to cry. "I'm so sorry, Kelly! God!" First he ran a few steps one way. Then he ran back. Finally he just collapsed next to the limp body. He pulled his knees up to his ears, hung his head low and sobbed. Big fat tears rolled off his face and plopped into the dirt.

LISTENING
Leslie Gee

Tara

I remember being frightened at night as a child, hearing things, seeing figures in the shadows of my bedroom closet, not being able to close my eyes and fall asleep, even when I was so tired I could barely keep them open. Each night, after the snores of my parents were loud enough to know that they would not wake easily to make me go back to my own room, I would tip toe down the hall and crawl in bed beside my mother. I was constantly assured that it was just my imagination, but I knew it wasn't. I swore to myself that when I grew up, I would listen to my children. I swore I would believe them.

That night though, I didn't listen. I didn't believe her. I told her she had just had a bad dream. I told her to go back to her own room. I tucked her in and kissed her forehead, turned out the light and walked away. I walked away even though I had just heard her give a scream full of terror that no four-year old could possibly understand. I was tired, I guess, tired of her coming into my room every night saying she saw something or heard something. Tired of my boyfriend rolling his eyes when she interrupted us.

Even weeks before it happened, days before we saw her there like that, she had told us. Her sister had told us she was acting strange, and I didn't listen. I thought it was another of her recent ploys to get my attention. I had known something was very wrong just weeks after I signed the lease, paid my greasy landlord and moved all our crap in. I felt something, something that was just wrong. Something gave me an uneasy feeling about my children, but I didn't know what. I was having dreams that would wake me up at night, but then I wouldn't remember them. My intuition was telling me something was wrong, but I didn't pay attention. I mean, Goddamn. We're city Indians for Christ's sake. I've been taught to tune that shit out.

That summer day in the laundry room, I had thought it was Robert standing beside me as I switched the clothes. I asked him to give me a hand, but when I looked, he wasn't there. He had been out in the car the whole time. I didn't say anything, and I don't know why. It probably started before that, before the night we heard her screaming in the hallway outside our bedroom...I saw it.

154

I saw her changing after that night, becoming less and less of herself until she was only that thing. It changed her. I would hear her in there just talking away, but no one was in there with her. Little kids do that though. Her sister told me that she saw her doing this or that, looking out the patio doors in the middle of the night, but that didn't surprise me. She was never afraid of the dark, ever. Even when she was just a toddler, she used to wake up in the middle of the night and play with her little toys in her room.

Rosie

That night when she screamed, it woke me up. It was the worst thing I ever heard before. She sounded so scared, and that got me scared. I didn't want to get up even to see what was happening. I just knew something terrible had happened. For some reason, I thought she had found mom and Robert dead in their room or something. It sounded that bad. I just covered my head with my blanket and waited. But nothing happened. It was Robert who got to her first. I listened to him, out there in the hallway with her, "Aubrey, what's wrong. Aubrey. It's okay. It was just a dream. Aubrey, why were you screaming?" She was still freaking out when he was talking to her. All I could make out of her cries was that she saw something move or something moved something. But, next thing, mom is bringing her back into bed and tucking her in. I asked mom what happened, and she just said that Aubrey had a bad dream. Aubrey told her though, she tried to tell her that she saw something. She was real scared and so was I. She was never like that before. She was just little and fearless, like a boy. She wasn't afraid of the dark-just walked around the house in the middle of the night like it wasn't nothing. She knew the steps to mom's door by heart. She went to sleep with her every night, but not because she was scared. So, that night, I let her crawl into bed with me since mom wouldn't let her. Later that night, Aubrey got real sick.

I'm the one that noticed, because I was laying there beside her. Her skin was real hot, almost burning. I woke up mom again, and she ran Aubrey a bath and rubbed alcohol all over her little body, but the fever wouldn't come down. Mom had to take her to the Indian Hospital in the middle of the night. When they came back, Aubrey was better, and she went straight to bed until the next night. She slept all day.

It was after that night that she started to change. I could tell. I told mom, too. I told her about Aubrey up at night. I told her how I found her there staring out the patio window at something and talking too. She was talking to something outside. A lot of nights after that, I would wake up at night and notice her out of her bed. She was always there, staring out the glass of the patio door in the living room, and sometimes she was talking too, but I couldn't make out any words. When I told mom, she said that she was probably just sleepwalking or something. It got me real scared, because I knew something was wrong. Only one time, mom asked her about it. Aubrey just said that she was talking to her friend, her little friend. Then she started doing other weird stuff, like being mean to our puppy and biting children, even babies at her daycare.

Robert

Things were happening at home before that. I hadn't mentioned it to Tara because she still gets so scared at night, just like a little girl, sleeping with a light on and all. I guess the first time I noticed anything was about a week or two after we first moved in. She got up late one night to go check on the girls. I was half-asleep, but I woke up when she got off the bed, and I asked her where she was going. When she came back, I swear I saw the figure of something following her back towards our room. She came back in and crawled back in bed and it moved in a different direction. I didn't want to scare her, so I didn't say anything. I just got up and walked out into the living room and looked around. I went back to check on the girls again, but saw nothing. I got back into bed beside her and left the door open. I was tired, but I know what I saw. I burned some sage that morning after they left. There were other things too. Things that were little, that could be explained by wind or electronic glitches, but I knew they weren't. I knew that Tara was spooked, so I made sure to act as if it was all nothing.

I guess it had been a few weeks since the first night we heard her screaming outside our door. It was still hot outside. I woke up in the middle of the night and walked into the kitchen to get a drink of water. I didn't notice her standing there. I tilted up the water jug and took a big long drink of ice cold water, propping the refrigerator door open with my foot as I drank to feel the cool rush of air. It was only when I was lowering the jug that I saw her tiny figure across the living room. It was dark. Only the dim light from the refrigerator lit the room. The sight of her little body there startled me, and I nearly dropped the water I was holding. Then, relieved that it was just her, I laughed a little and called to her, "Aubrey." She didn't answer though. She didn't say a word or move. She just stood there in front of the patio door, looking out. "Aubrey." I called to her again, but she didn't move. I noticed something in her hand. It was dark, but I saw something there hanging, almost touching the floor from her hand. I didn't think twice about it. I walked toward her. I thought that maybe she was just sleep-walking or something. "Aubrey. Aubrey, wake up." When I moved closer, I saw a puddle of something wet on the floor. It was leading to the limp thing hanging from her little hand. I got frightened and called to her in a stern voice, "Aubrey!"

Tara

It was Robert yelling her name that woke me up. I jumped out of bed and ran into the living room. I saw him standing beside her. She was turned the other way facing the patio door. Just then, he touched her little shoulder, and she turned around. Her face, I can't even describe it. It was her face, but it was changed into something, something evil looking and ugly. I must have screamed. When Robert saw it, he tried to step back away from her but slipped in the puddle beside her.

Rosie

I heard Robert and mom in the living room the night it happened. I heard mom scream, then nothing. When I got into the living room, Aubrey was gone. Robert had blood all over the side of his body and was running outside after her I guess. Mom was just standing there all white and crying. She yelled at me to come here. I don't know where my little sister went. Robert never found her, just our

157

puppy's body. It's belly was open, and nothing was inside. I don't know though, because they never let me see it. That's only what I heard Robert tell mom.

Now, we're just waiting I guess. We're just waiting to see what will happen next. We can't leave though. We can't leave our house because this is the only place my sister knows. It's cold now. At night, I sleep with mom and Robert. I wake up in the middle of the night, though, and I listen. I listen for her little voice outside the patio doors. I listen for her because we miss her, because she still might come back.

Officer Padilla

When I got the call, I figured it was for domestic violence. The neighbor said she had heard screaming, then nothing. When I got there, though, no one would answer the door. The front window was open, and I could hear crying. So, I called for back up and kicked in the front door. I saw the little girl first. She was standing there in front of their back patio door, crying. She was so little. There was blood all over her. I knew she was in shock because she wasn't talking. I searched the house. That's when I found the other three bodies. They were still in bed, all three of them together. They were murdered in their sleep. It's a miracle this little one survived.

BLUE GLASS
Jamie Natonabah

One day, he began to see the back of a woman's head instead of his face in the mirror. When he leaned forward, the woman's head leaned forward too and he could see the dirt in her brown hair. He noticed the tangles and split ends when he moved his head from side to side. He did not leave his house for two days. Only after he'd worn down his fingers halfway through the fingernails did he wonder if it was his mirror that was broken. He was tired of trying to reach through to the other side to the reflection.

Very early in the morning he placed mismatched shoes on his feet and opened his front door to leave. He discovered the illness was not confined to his bathroom mirror. He walked the streets and tried to remain calm as he watched his reflection in shop windows. It was the woman with her head turned away from him. He began to walk faster; paying no attention to people he shoved out of his way. The woman walked faster too.

He laughed as he began to run, watching the woman's matted hair flap behind her as she began to run too. He leaped off the sidewalk laughing hysterically. The people on the bus that hit him will always remember the sound of his laugh, like blue glass, broken in his mouth.

HUMMING
Sara Marie Ortiz

It's what she wanted. It's what she wants. That's what I kept telling them. I told them, but they were done listening to me. One day. If you wore my shoes, walked where I walked...you'd probably be... like her, like me. Let me tell you my side...just let me tell you. If you had to see her dirty face and smell her cheap vodka dipped stench everyday, you would have lost your fucking mind...long before I did. It won't kill you to hear a lost man's story. Stay, because, you don't know the worst of it.

It was the humming that finally got me. Her song made me sick. Always the same. Always that same ugly mouth humming. When she hummed, she prayed. I heard her. No one else could hear her. But I did.

I never thought I could do it. I saw her day after day after day. I moved to the block seven months ago. Seven months ago on the fourteenth. She dragged her feet past my door that first evening after I'd arrived. I was unpacking the few boxes of crap I'd brought from the apartment I'd been sharing with Renee. Some of it was hers; most of it was her hangers, a few faded pictures of us, and the crock-pot we had bought at the flea market on the day we moved to this place. "The Land of Enchantment." The travel books lied. We flipped through book after book trying to decide. Renee said that she wanted to be in a place that reminded her of home...this time without the voices. She said that she would miss the ocean and the lights from the hill, but the angels were tired like her now, and she needed to remember what it was like to breathe again.

I didn't even notice her at first. She was dirty. Man, she was dirty. Her hair all in clumps. Beneath the layer of dirt I could tell she was not very old. I would see her through the bars on my windows, through the bars on my door, and I could almost touch her hurt. Seeing her made me sick, but I didn't have much of a choice. She would sometimes look up at me as she dragged her heavy body past and it looked as if the pain of raising her head to meet my gaze was almost too much. Her eyes told me stories that I didn't want to know. What she carried around made the both of us embarrassed. And day after day she would come to remind us not to forget what she carried.

Central Boulevard. I moved into the Desert Sands Motel on the day
Renee left. I told myself it was only temporary. One brown and yellow
stain on the ceiling, check. One rusted and dripping faucet, check.
One bullet hole through the motel room window, check. The hot days
were my favorite. I told myself that it could be worse. I told myself
that one-day soon, she would come back to me.

I watched the dirty woman get beat by guys with dirty brown faces like
hers. I saw her yelled at and chased from restaurants and gas stations
by men with hot, red faces that turned soft and filmed with their guilt,
as they watched her walk away. I saw the woman eating food from
dumpsters and cans along our street, and sometimes she would catch
me watching her. Most times she would get real embarrassed and
stop, but sometimes she would just keep stuffing fist-fulls of smashed
and crumbling whatever into her filth rimmed mouth, turning her back
to me if I watched too long.

The sun was blazing that day. The evening was coming, and by that
time things had usually cooled down. The really needy came all night
long, but most usually gave me a break for dinner. Not that day.
Their eyes, their faces, their hands. On that day, it was like they didn't
know how to do anything but reach. Tendrils of hot pain moved
beneath my eyes, burning hotter each time I reached into my pocket.
Maricela came around three. The bitch brought her goddamn kids
again. The baby started screaming just as they were walking up, as
soon as she saw me. The little boy, covered in chocolate something,
ran up and wrapped his coated hands tight around my knees. I had
only known Maricela for four months. Her kids saw me, and they
knew me. I hated those fucking kids then. Manuelito, her little boy,
could have cared less. He just stared up at me with those wet brown
eyes as the little yellow stone passed from my hand to his mommy's.

After they had left I sat down on the porch and lit joint number 13.
I sat outside my door, watching. The woman was there, pawing
through the dumpster behind the Long John Silvers. I watched as she
pulled a half-squished tomato from inside. I watched as she brought
the thing to her hungry mouth. The seeds and juice dripping down her
dirt caked chin onto her dingy, yellow-white sundress.

I watched with my eyes and mouth wide open in disgust as she devoured the dripping fruit and as she licked her fingers clean; her shiny pink tongue darting and flipping out and in, in the late afternoon sun.

I felt it as it happened. It was four eleven. She smiled and then began to hum. She smiled. It was time.

I was never so ready in my entire life. I never felt so alive. Suddenly everything was so clear; it all made so much sense. I went inside my room and began to get things ready. I pulled out a bottle of my finest Hennessy and set it on the nightstand. I rolled three joints and set them on the T.V. I dug through my bags, searching for a tool that was just right. When my fingers brushed across the cool skin of my baby, my chrome glock, I almost rethought the whole thing. I almost took her to the desert. That's when I saw it. The jagged edged blade that Neto had given me for my fourteenth birthday. It was perfect.

He had handed it to me and said, "For shanking those ones that don't know how to stop, I know you'll use it well little vato, don't tell your mom."

I slipped the blade from its black and red leather sheath. I had never felt so alive. I looked down at the blue roads on the underside of my arm as I held the huge knife. It felt so right. It was heavy in my hand. It was begging me to take its cold weight to a warm place. I fingered the edge. It was so smooth. I placed the blade gently back into the sheath and jumped in the shower. I dressed in my black County pants, a clean white T-shirt with the creases ironed perfectly, and my black lambskin coat. My chain with the SUR emblem hung from the dresser mirror. I placed it so carefully around my neck. So carefully. I touched every link and rubbed the shine of the metal emblem for the last time. I closed the blinds as I left. My fingers shook.

I remember each step I took. The sound of my blue-sueded Stacey Adams tracing the steps of an invisible Death Row, across the broken stones, of the desert pavement. By now the sun had relented and the sky was the color of flesh roasting over burning coals.

The woman was gone from the dumpster.

I walked a bit up the street past the·Long John Silvers to the Sex Toy Warehouse, the Castle Superstore. The light from the white sign, with the big blue castle and "royal" red letters, was glaring. I saw my reflection in the glass front, painted white on the inside to conceal the toys and naughty girls and boys within. I caught the glint in my eye and was surprised for a moment. I had not seen that sort of light in many days. The glare from the superstore shone against my chain, the light bounced from it and into my eyes and I smiled for the first time in months.

I hadn't walked four blocks when I saw her. She was a few steps down Lemon Street. Standing in front of a pink stuccoed house with a chain link fence. She was reaching over and rubbing the scruff and behind the ears of a brown scab and tic covered Pit. She rubbed him with both hands. He nuzzled into her palms and licked her fingers with his big flappy tongue. She laughed gently and kneeled down to the chain link fence. The dog pushed his nose up to the links and the woman whispered, "Soon. Soon, boy, you'll see."

I walked up to her then. She was not startled at all. She almost looked as if she was expecting me. My voice almost cracked, but when I said it, she heard me right. "Do you want to come with me?"

She said nothing. She was still kneeling, and she spoke not one word as she rose and walked silently past me to where I had stood watching her moments before. She stopped and turned. Her eyes burned into mine for an only a moment before she disappeared around the corner. The sky was now cast with shadow and the avenue was coated with a dull evening light. I walked until my steps were with hers. Her breath came slow and steady. Mine was raspy and fast. We didn't speak almost the whole way to the room. We just listened to each other, breathe.

As we approached the parking lot of the motel, she stopped. She looked to the West, the tiniest jewel of fire burning in the sky above the west mesa.

I was shaking as I turned the knob. I could hear her behind me, and a wave of what can only be described as sickness coated with sadness filled me. She entered slowly, looking around the room, as if she was taking notes, trying to remember everything. I closed the door, and turned to her as she sat on the edge of the unmade bed.

"Do you want a drink?"

"Sure."

I reached for the bottle of cognac, broke the seal, and handed it to her, watching closely as she raised it to her dirty lips. She drank with large thirsty gulps, as if the stuff was the purest water from the purest spring. And I just watched. She handed me the bottle, and I set it on the dresser. And what happened next, I will never understand. I knelt in front of her and put my head on her knee. She began to stroke the back of my neck with soft careful movements. It was as if, she knew, and I knew, that this wasn't the way things were supposed to go, but would anyway, because we were both just so alone in the darkened concaves of this city. I raised her dress, and she laid back without a word.

I began slowly. She was trembling. I grabbed her by her hips and pulled her closer to me. I was surprised, since from the look of her, I wouldn't have ever thought that her pussy would smell like home. A mix of cedar, honey, and something smoky I couldn't quite recall. As hungry children do, I suckled her soft, wet skin, and I was almost gone, when, she began to hum. Like a person startled from a dream, I moved to the dresser, took the shiny metal from its sheath. She was so peaceful there, amidst the ruffled white of the hotel sheets. She was still within her dream, and she continued humming. So softly. Her eyes were closed as I took the cold blade to her trembling throat.

Days and nights passed. Clouds, and rain. Wind and sunrises, all around me, all inside of that tiny motel room. Really, it was dark the whole time, and it only took a few minutes, for her final struggle to end.

Peaches. I've always loved them. Summertime. When the food stamps would come. Mama would buy just the things we needed, but she knew how much I loved peaches. I would peel back the soft skin, so carefully, savoring every last bit of the juicy fruit. Mama would let me cut them myself.

164

She was a peach. But lying there in the motel bed, the white, white sheets now completely red, she wasn't quite right. I had an idea. I turned the bath faucet slowly, turning the hot water all the way to the left. I picked up the bar of soap, and washed my own hands carefully and slowly, dried them, and hung the white hand towel back over the edge of the bathtub. I study my hands, even now. I think they are some of the nicest hands I've ever seen. My mama would say, "You have my hands, mijo. Even when I am gone. You can always look at your hands, and you will find me there."

I was scared. It's hard to say it now, but I hesitated before I placed my right hand behind her neck, and my left in the crook of her knee to lift her from the blood soaked bed. I expected her body to be completely limp, but it wasn't. I expected her skin to already be cold, but it wasn't. I caught the misted reflection of her and I, as I stepped past the mirror, foggy from the bathtub's steam. A flash. A memory. A sacrifice. An Aztec prince. An Indio` god with a gleaming, and feathered headdress, holding a woman in his arms, dripping. He has just saved her, and all of Maztlan hails this one warrior.

I didn't close my eyes. God, I wanted to. But I couldn't. I couldn't miss even a moment. I could feel my lids heavy, resisting my will to keep them open for the task. She was so beautiful, then, and I hadn't wanted to know it, hold it, until that moment in the motel bathroom. I laid her so gently down against the white of the tub. I touched her lips, still warm. And, I thought I could actually feel it slip under my touch. The water was just right, steam billowing, the water flowing hot from the faucet, and I washed her face with the white washcloth, as the color drained from her cheeks. I reached between her legs, soap in hand.

I lathered up the soap real good. From head to toe, I lathered, and rinsed. Lathered and rinsed. But the blood kept coming. The water was tomato soup. The bubbles weren't strong enough. I kept wanting her to open her eyes, so she could see what she had made me do. I stood up, looked into the mirror, deep into my own brown eyes, and at the drops of blood splattered on my cheeks and chin.

There was blood under my fingernails, covering my hands, up to my wrists. My white T-shirt was not white anymore, and I smiled at the red shape her head had left on my sleeve when it slumped to the side as I was reaching to wash her hands.

She wasn't clean yet. She had given herself to me. I left her then. I waited until the bleeding stopped. I smoked a joint and began washing her again, taking special care with her neck and face. I pulled her hair back, and caressed the spot just where the cut ended. I thought maybe I loved her. Or perhaps not her, but rather, the way she had come and gone, out of this brown body that I washed, without ever saying anything displeasing.

And now, the only problem was that I wanted to her to see what I had done. I wanted her to speak to me, and tell me how grateful she was that I had come when I did. She called me with her humming. Her song told of where she had been, and where she was going. Though even now, I still can't say where that is.

Maybe you're not like me at all. Maybe you're nothing like us.

They won't let me have my own soap you know, and now, my hands are always dirty. I keep looking, and looking at them. Mama said she would come, but she never did.

DRAMA

THE SWIM SUIT
Jody Barnes

The spot light comes up on an elderly lady sitting in a wingback chair. She is wearing a house dress and sturdy black shoes with hose. She speaks over the tops of the audience's heads, almost as if she's blind.

Old Lady

I was fifteen going on sixteen and my friends and I spent long afternoons at the beach, watching the boys and giggling in the hot sand. I loved the gritty feel of it as I wriggled my toes deeper and deeper. We'd walk up and down the boardwalk for hours while old men in their shirt sleeves suspenders fished from the rails.

(Pause)

One day, about two weeks before my birthday, I was out shopping with my mother. She needed to get some new hose and a...

(She holds up a hand and whispers conspiratorially to the audience)

A new garter belt.

(She chuckles)

My mother, was so old fashion, she couldn't even say GARTER BELT! As we walked up to Spurgins Fine Women's Wear, I noticed one of the mannequins had on the most gorgeous swimsuit. It looked like something Ava Gardner would wear. It looked like a sleek little toga; it had little straps that plunged into a crisscross in the back. High cut legs. "Mother look!" I said, "That's what I want for my birthday! Oh, please?!" But she pressed her lips together in disapproval. I also knew it was really my father who would make the choice and he definitely wouldn't like it. But I had to have it, it was such a fine shimmering thing, I didn't even care whether or not it fit me, I just couldn't allow anyone else to posses it. I was determined I would have it. I begged my parents day and night.

"No." My father said. And my mother, she just followed my father, like always. I pleaded, I cried, I reasoned, I told them if I could have the suit I'd wash dishes every night for a year. My mother said if I kept on about it, I'd be doing dishes every night for two years. And my father's face remained a stone.

Finally the day of my birthday came. I tried to pout, but it was so nice. Mother made my favorite meal, roast beef with her delicious gravy, mashed potatoes and green peas; we had coke-a-colas. After everyone was done eating, mother brought out the cake, yellow, with chocolate frosting. My name in big pink script. Then came the presents, my little sister gave me hair combs and my brother gave me a beautiful stationary set with matching envelopes. I remember he blushed deep red when I hugged him.

(She gives a twittering laugh)

Then came the present from my parents. When I tore the paper off and in big letters red letters 'Spurgins Fine Women's Wear on the box and I knew, I knew. I started jumping up and down and screaming! I threw open the box and there wrapped in red tissue paper was my swim suit! I hugged my father. I said, "Oh daddy, daddy, thank you!"

(Her eyes focus in on someone in the audience laughing and she speaks to them)

And he said to me. "You are a young lady now; I suppose I should let you dress like one."
When my mother kissed me she had tears in her eyes.

(She goes back to speaking over the heads of the audience)

The next day when I wore my new suit my friends all gathered round, squealing with delight and telling me how lucky I was, that my new suit was to die for! Except Susan O'Malley. She came from one of those big Irish Catholic families – she was second youngest, so everything was a hand-me-down. Her suit gripped in the front and was nubby in the butt. I felt like I was walking 3 feet off the ground when I wore that suit, it was so shiny and sleek. The boys would turn to watch me walk by; I would pretend not to notice.

(She picks up her cup of tea on a saucer to take a drink. But the *delicate cup starts to rattle on the saucer so she quickly puts it down)*

One night, there was a knock at the door and there were men in dark suits with papers that said we no longer had any legal right to live in our house, or for my father to own a business, or work, or for us to go to school.

(She frowns, again finds someone in the crowd and speaks directly to them)
They told us we had fifteen minutes to gather what we could carry. Fifteen minutes.

(She shrugs her shoulders, looks away)

I ran upstairs and as I passed my mother's room I saw her sorting through her glove drawer, trying to find the right gloves to wear, while my sister laid across the foot of the bed crying. In my room I pulled my suitcase from under my bed and threw in some underwear and socks, a couple of sweaters, and then I saw it, hung from my bedpost, my swimsuit. I quickly stripped off all my clothes and put on the suit, then put my clothes back on over it. Then I knew it was time to go because they were pounding on the door...out fifteen minutes were up.

There were more families like us out on the street, their suitcases and their bundles and their things they couldn't bear to leave behind. And there were soldiers.

(She leans forward, looks into audience)

They herded us like animals through the streets of the town that was once my home. Past people who were once out neighbors and now didn't lift a finger to help us. I saw one of the watchers throw a rock and it hit an old man in the head and knocked him down into the mud, blood trickling down his forehead. The soldiers poked him with their guns until someone helped him up and away.

(She sits back, sighs)

They marched us a long way that night, past the dump, past the stock yards, past the out laying farms, they marched us through the night until there was just enough light so we could see the monstrosity they were marching us to. There are so many horrible words for what that place was...reservation, stockade, prison, gulag, concentration camp, internment camp...they all mean the same thing, really.

They marched us through the gate and then split us up, men and boys one way, women and girls the other. They led us to a huge concrete building and told us to strip naked. If we had any valuables they were to be left, anyone trying to sneak anything would be punished. The women stood staring at the floor; these were modest women, shy women. Women who've probably only been naked in front of their husbands. The prison guard went up to a woman and ripped the dress from her. Then the guard yelled, "Ladies, don't make me get the dogs!" and just like that, everyone unfroze. Fingers found buttons, zippers were tugged down, bras were unsnapped.

(She looks down at her hands in her lap)

That was the first time I ever saw my mother naked. She looked so small.

(She looks out into the audience)

The bones of her spine when she bent to take off her stocking so sharp and delicate. Her shoulder blades like stunted wings.

(She smiles sadly)

I stripped down to the swimsuit, but then I stopped. I stood there crying, thinking about Susan O'Malley and all the boys. I couldn't leave my beautiful suit behind, I wouldn't do it, but then I heard the dogs outside and I let the swimsuit drop to the floor. As they herded us out the door I craned my neck back to see if I could catch a glimpse. I couldn't.

They sprayed us down with Lysol, scrubbed us with rough brushes until we bled. They chopped our hair odd. They stored us in these so called shelters that provided no shelter from the wind or cold. They fed us maggoty food and dirty water that caused us to die of dysentery. They sent us out on work crews that people never came back from.

I lost my mother and my sister in the first 3 months. I don't know what happened to my father or brother. But once, just once I thought I caught sight of my father. He was in the back of a work detail truck. He was wearing a white shirt and tied around his upper arm was a piece of black cloth. It looked so familiar...

(She closes her eyes and reaches out a hand, rubs her fingers together)

I wonder, if I could have touched that cloth...

Monica and Oscar enter a doctor's office. The room is quiet, yet full of people. They take a seat in a corner where Monica seems anxious. The only sounds are light sniffling and occasional coughing. Monica looks around and then breaks the silence.

                    MONICA
        So you really wouldn't try?

                    OSCAR
        No, that's disgusting.

                    MONICA
        But wouldn't you even be a little curious? I mean, how
        couldn't you?

                    OSCAR
        Because I'm not sick and twisted like you.

                    MONICA
        Oh, come on. That's not fair.

                    OSCAR
        Okay, first off why would you even think about doing
        that?

                    MONICA
        It's not like I sit and think about it all the time...I mean,
        it's just one of those things.

                    OSCAR
        But it's still not normal. It's still a twisted thing to just
        randomly come up with when you're bored. Sick,
        sick, sick.

An old man next to Monica coughs loudly as if to ask them to keep it down.

                    MONICA
        Oh please, don't tell me that you're beyond strange

thoughts. That all you ever think about is your work load, dinner, what movie you'll see next, etc.

                    OSCAR
Okay, maybe you're right about that, but that doesn't mean that I have to spend my time contemplating things like that. I mean, do you sit around wondering how you would kill someone if you were doing it for fun?

The old man looks slowly over to the two. He is slightly disturbed. When he sees them looking back, he quickly looks ahead.

                    MONICA

Oh, come on, it's not like that. (to old man) it's really not like that. (back to Oscar) Regardless of why I thought of it, I still did and I don't know what you're so afraid of. Why is it such an awkward subject?

                    OSCAR
It's awkward because it's creepy.

                    MONICA
So if you were just hanging out at a friend's house and they offered it to you, you wouldn't be a little curious and want to try it out? Wouldn't you be the least bit tempted? Come on, I personally think it's normal. I know people don't like to talk about this kind of thing, but everyone has natural urges.

                    OSCAR
Honestly? If I was at a friend's house and I had the chance offered, I'd have to seriously reconsider our friendship and then take a long time by myself to ponder my error in judgment.

                    MONICA
That's beside the point. You still haven't told me if

you'd try it. Okay, let's just pretend that it isn't sick and twisted, would you try it?

                    OSCAR
Well, of course I would because it's no longer something disgusting.

                    MONICA
Why are you being so irritating? Fine. Since you can't be open minded enough to even talk about something, I'll let it go. Okay?

                    OSCAR
Okay, fine.

They sit and scan the room. Faces of others in the waiting room look both curious and horrified. Monica, despite her settlement seems anxious still.

                    MONICA
Oh come on. You know this drives me crazy! If it's no, then say no.

There is silence for a moment. The old man turns to look as if he too is anxious for the answer.

                    OSCAR
Okay, fine. I wouldn't say I'd never do it, but there are certain things to take into consideration.

                    MONICA
Like what?

                    OSCAR
I mean, I wouldn't just- (he becomes very aware of his surroundings) just...

                    MONICA
Just what?

OSCAR

Just...do it. I mean, I would want a full background check.

MONICA

What? What for?

OSCAR

You know...make sure they're clean. I wouldn't want to pick up any diseases or put myself in jeopardy simply because someone else decided to be careless. There are a lot of disea- things, things to worry about introducing into your own body.

The old man who has been watching slyly up until now turns away in disgust.

MONICA

So for example, you would take Mother Theresa over...let's say Drew Barrymore?

OSCAR

Oh yeah, of course, any day. Drew Barrymore? That's just a bad choice. How could enjoy that? At least with Mother Theresa you know what you're getting. Well, of course there is the age factor. Who would really want that? I mean, she's old and wrinkly. You know, I retract my statement. You're on the right track at least. Mother Theresa was a great woman who did a lot of great things, but this is me we're talking about. I like mine young and fresh.

MONICA

Well, that's not fair. If we're taking personality into affect, that's different. Is that what you're saying? That personality plays into this?

OSCAR

Just as far as personality affects personal hygiene. I

mean, you can bet that Mother Theresa wasn't a party girl. Who knows what all Drew Barrymore has. That girl is dirty.

MONICA
But what about animals?

OSCAR
What do you mean?

MONICA
You never cared before about the sleeping habits of cows you've had. Why does it make a difference here?

OSCAR
Well, you have a point there. But humans have a greater ability to put themselves in harmful situations. A cow's life is pastoral, on the farm, you know...simple. I like simple. I would definitely trust a cow over Drew Barrymore.

MONICA
Well of course you'd choose a cow over Drew Barrymore, I mean, shoot so would I. But just comparing - let's say within species. How do you know you've never had a sickly cow? I mean with all of the hormones they stuff those things with, how do you know you're safe? How do you know you haven't contracted anything from one of them?

Monica and Oscar become conscious of their volume as the others in the waiting room appear uncomfortable.

OSCAR
(lower) Look, you're right. You know me, so that means I've had God knows how many cows. Well, when it comes to matters of necessities, I prefer don't ask don't tell. I don't need humans.

                        MONICA
I don't know. I'd be too curious.

                        OSCAR
        You'd do it?

                        MONICA
Of course, and I wouldn't need a background check
either. It'd be all too tempting. I'd have to know. I'd
just have to have it, at least once.

                    LOUDSPEAKER
Mr. Montoya? Oscar Montoya, the doctor will see you
now.

                        OSCAR
        I'll be right back.

Oscar stands and heads over to the door leading to the doctor's
office.

NON-FICTION

I have to say that before you read this, you may get angry.

In the first part of this series, I focused on a presentation given at the Institute of American Indian Arts by Dr. Peter DeBenedittis called "Seduce Me" in which DeBenedittis talked about how the media affects our lives and subsequently our decisions. While DeBenedittis' presentation focused mainly on the changes in culture that have occurred as a direct result of media influence, such as its effects on our self-image and our perceptions of others, this second part will focus on our decision-making process as a nation.

The more and more that I read and the more I investigated, the more disgusting the facts got.

When I think about the mothers that will never have children because chemicals from GE got into their bodies, it saddens me. When I think of rainforests being destroyed and our resources being literally devoured by corporations, I wonder when it will stop. We have faces and personalities, and we have our own problems, too, but nobody deserves to have their lives ruined or destroyed so someone can make a little bit of money.

Maybe people just don't want to start looking for solutions until it happens to them. Maybe when you, or I, realize that we will never have children because our organs have been permanently destroyed or mutated, we will get the message.

Mothers no longer have the ability to produce children, the greatest gift anyone can have, and we have not heard anything about it because the same corporation that took that gift, owns the media outlets that should be reporting about it. Maybe when our children can no longer drink fresh water or see more than archival pictures of forests, we will realize something is wrong. When we cannot eat food that has not been genetically modified, we will see that something has gone awry. When we find that nearly every aspect of our lives has been marketed and taken from us, maybe then we will take and stand and say, enough.

Perhaps our lack of action or realization is our inherent selfishness, or perhaps it's what we have been taught. Either way, one day, one of us, either me or you who are reading this, will be a victim of this corporate recklessness, and all I can say is, I'm sorry. My heart aches for you just thinking it might happen. I'm sorry.

To pick up a little from where we left off last time, the following corporations own each one of these media outlets:

1. AOL—Time–Warner: HBO, Cinemax, Comedy Central, Time Inc, Time Magazine, Sports Illustrated, People, Entertainment Weekly, Fortune, Money, Business 2.0, Southern Living, Popular Science, Outdoor Life, Field and Stream, Parenting, Family Life + (43 other magazines), Time – Warner Trade Publishing, IPublish.com, Little, Brown and Company, Warner Books, Warner Music (Roger Ames), Atlantic Records, Elektra Entertainment, London-Sire Records, Rhino Entertainment, Warner Brothers Records, Columbia House, Maverick Records, RuffNation Records, Strictly Rhythm Records, Sub Pop Records, Tommy Boy Records + (10 other recording companies), Warner Brothers Studios, Warner Brothers Pictures, Warner Brothers Television, Warner Brothers Animation, Looney Tunes, Hanna-Barbara, Castle Rock Entertainment, Telepictures Productions, Warner Home Video, MAD Magazine, DC Comics, New Line Cinema, Fine Line Features+ (5 other studio production companies), Time-Warner Cable, Turner Broadcasting, The WB! Television Network, CNN, TBS Superstation, Turner Network Television, Cartoon Network, Turner Classic Movies, Court TV, Atlanta Braves, Atlanta Hawks, Atlanta Thrashers, America On-Line, Compuserve, Digital City, Digital Marketing, ICQ, IPlanet, Mapquest, Moviefone, Netscape.

2. Walt Disney Company: Walt Disney Studios, Walt Disney Pictures, Touchstone Pictures, Hollywood Pictures, Caravan Pictures, Capital Cities, ABC, ABC Television Network, ABC World News Tonight, ABC Family Channel, Fox Family Channel, Saban Entertainment, Disney Channel, The Soap Network, Toon Disney, ESPN, Lifetime Television, A & E Networks, The History Channel, The Biography Channel, The

Style Channel, E! Entertainment, Fairchild Publications, Chilton
Publications, Diversified Publishing Group, Miramax Films, Walt
Disney Television, Touchstone Television, Buena Vista Television,
Go.com, Infoseek.

3. CBS – Viacom: Paramount Pictures, Paramount Home
Entertainment, CBS, CBS Television Network, Special Events
Coverage, CBS Evening News With Dan Rather, CBS Early Show, CBS
Radio, Infinity Broadcasting, Group W, Country Music Television,
Nashville Network, Viacom Television Stations Group, Paramount
Television, United Paramount Network (UPN), United Cinemas
International, Famous Players, Blockbuster, Paramount Parks,
Showtime, MTV, The Music Factory - Netherlands, MTV Dance –
Britain, MTV Live – Scandinavia, Cecchi Gori Communications – Italy,
Nickelodeon, VH-1, TV Land, Noggin, Comedy Central (jointly
owned), The Movie Channel, Country Music Television, Flix, The
Sundance Channel (jointly owned), Simon and Schuster, Scribner, The
Free Press, Pocket Books, Spelling Productions, Famous Music.

4. Vivendi: Seagrams Gin Company, Ltd, Universal Studios, Universal
Music, Farmclub.com, Interscope Geffen A&M Records, Island Def
Jam Records, MCA Records, Motown Records, Mercury Nashville,
Verve Music Group, Lost Highway Records, PolyGram Records,
Deutsche Grammophon, Decca-London, Philips, Computer Games,
Blizzard Entertainment, Sierra, Universal Interactive, Flipside Network.

5. News Corporation (Rupert Murdoch): News Corp Publishing,
Harper Collins, William Morrow and Company, Avon Books, Amistad
Press, Fourth Estate, Fox Cable, FX, Los Angeles Dodgers, National
Geographic Channel, Fox Magazines, The Weekly Standard, Fox
Television Network, BskyB, Channel(V), Sky Perfect TV, STAR, Stream,
20th Century Fox Films, Blue Sky Studios, Fox 2000, Fox
Entertainment, New York Post, TV Guide, Festival Records, Mushroom
Records, [16 Australian and 9 British newspapers as well].

6. General Electric: NBC, Dateline NBC, NBC Entertainment, Today
Show, NBC Nightly News, Columbia Tri-Star Pictures, RCA

The idea that government, corporations, and media are all intrinsically intertwined is not new, yet why do so few believe it, or for that matter look into it for themselves? After some intensive searching what I found was very disheartening, if not Orwellian in nature.

General Electric's Political Connections

To pick on General Electric first: In 2002, GE received a defense contract worth $1.9 billion from the U.S. Department of Navy. The contract was for the purchase of 480 F414-GE-400 engines used in the air fighter plane known as the Blacker and in F/A-18E/F Super Hornet Strike Fighters, and an order for 13 spares and modules for the U.S. Navy F/A-18E/F Super Hornet Strike Fighter.

While GE specializes in media outlets, they also specialize in weapons parts. And while this would make GE a prime candidate for developing the engines for the Super Hornet Strike Fighter and Blacker, it may have also had a slight inside track from campaign contributions.

According to the GE Workers United website, GE contributions can be broken down by presidential, U.S. Senate, and U.S. House of Representative races: "GE PAC contributed $268,500 to House Democratic candidates and $357,900 to House Republican candidates, $100,600 to Senate Democratic candidates and $147,500 to Senate Republican candidates. In addition, GE gave $5,000 to the George Bush campaign. The Company also put up $100,000 for the elaborate inauguration ceremonies for the new president. No money was given to Democratic presidential candidates."

But the weapons that GE constructs not only have an effect on the people that are assaulted by them, it also has an adverse effect on the people and environment here at home.

General Electric, No Friend to Humanity

In 1998, GE was ordered to pay $200 million in damages for

its pollution of the Housatonic River in Massachusetts. The payment was for GE's use and disposal of polychlorinated biphenyls (PCB's) from their plant in Pittsfield, now a holding area for toxic chemicals and river sediment drudged from the Housatonic. The closed-down plant, incidentally, sits within 50 feet of an elementary school. PCB's have been known to cause immunal changes, behavioral alterations, and impaired reproduction, and to affect both humans and animals, bind strongly to soil, and travel long distances in air and water. PCB's are found in plant and animal life forms that have been exposed to them and later in humans who eat the plants and animals.

Most tests to find out what effect PCB's has on children were conducted on mothers who had eaten fish contaminated with PCB's. At the time, the lawsuit was pushed by the Environmental Protection Agency (EPA), but today, GE may not have to worry about such strict rules with Bush's new appointments to such government positions like that of the EPA.

Upon entering office, the Bush administration nominated Linda Fisher to be deputy director of the EPA; unfortunately (or maybe fortunately), she only became deputy administrator. If you are not familiar with Fisher, you may be familiar with the chemical company, Monsanto, which has been fingered by the EPA for being the responsible party for the contamination of no less than 93 sites. Fisher used to lobby for the company.

The Media Link

While Fisher and Monsanto don't necessarily have anything to do with the media, Bush's other appointments do. The fourteen-member panel of the commission for the reform of Social Security includes Richard Parsons, co-chief operating officer of AOL-Time Warner. Secretary of State Colin Powell was a former board member of America Online, while his son Michael, head of the Federal Communications Commission, in 2000, finally voted to approve the merger between AOL and Time Warner, the largest media merger in history.

While this may not be totally shocking, one could sum up that the media outlets that Time Warner owned finally merged with the Internet technology that AOL owned. When the Internet, promoted as being the thing that would break up industry and media monopoly, is more or less dominated by one company with a distinct hold on other major media outlets, it creates a shrinking effect in cyberspace.

The Internet, therefore, has not necessarily failed to break up the power elite; it just seems that the power elite found a loophole to take it back. As a result of the merger, 2,400 employees were axed from the two companies, yet in corporate America, poor labor relations are always just around the corner, and in some of the happiest places on earth.

Disney's Sweat Shops

For decades, Disney has marketed to children around the world with such characters as Mickey Mouse, Donald Duck, and newer characters, such as Aladdin and Pocahontas. Yet while children seem to be the target audience, they along with women are also used as slaves to make Disney's products.

For the last eight years Disney has sold shirts for $17.99 a piece. The shirts only cost them fifteen cents each to produce. At least that's what the workers got paid per shirt. Workers in Bangladesh were forced to work for Disney for fifteen hours a day, seven days a week and then were beaten. Once workers organized for their rights, Disney pulled its work from the factory and dumped the women on the street. Disney has still not returned to Bangladesh.

Ongoing Exploitation

Children in Haiti can expect to make seven cents per pair of Pocahontas pajamas that they sew. Disney then sells them at a price well over ten dollars. And don't even get started on the Pocahontas thing either. It's bad enough that we have to have the word, American, tacked onto any set of words used to describe us like American Indian or Native American after being nearly wiped out by Americans, but then to have new generations of Americans try to re-write their history just to feel better about their crimes is even worse. So they killed us off, made movies about us that weren't true, and made children in Third World countries work as slaves to make the products to sell to the descendents of the people who killed us off.

It's kind of sick in a historical context; the moral sense is something totally different. Yet Disney's response to sweatshop work has been far from satisfactory. After Kathy Lee Gifford, from "Live with Regis and Kathy Lee" on ABC, suddenly discovered her clothing line was made in sweatshops, ABC mounted a campaign to save her image. If you missed the above, ABC is owned by Disney.

ABC's "Primetime" attempted to bolster Gifford's image by focusing on her charity work. No doubt Disney would love to change labor laws here to continue their child exploitation. Unfortunately, when it comes to changing laws to fit corporate needs, Disney is not as lucky as Viacom.

The Media Control of Viacom

Once Viacom acquired CBS, it was faced with federal law stating that no corporation can own two national television networks. Since Viacom owned UPN also, it was faced with the tough laws of the FCC. Luckily for Viacom, after a little lobbying the FCC decided to help the media giant out and change federal law. However, the merger didn't only change the laws, it raised some interesting issues.

In 1999, when the merger was announced, Viacom was poised to become the largest media conglomerate in the world as AOL and Time Warner had not yet jumped in bed together, and lots of people were very nervous.

In a Society of Professional Journalists paper examining the merger, professor and reporter Edwin Diamond was quoted as saying: "TV news has become too money-making to be left to news people. More and more, Wall Street sits in the executive producer's chair . . .When news went public and had to answer to stockholders with higher profits, the picture changed and newsroom budgets began to be squeezed to increase profits. . . . The result is often passive news, soft news, personality news, crime news, and news as entertainment."

Yet viewpoints on why this merger might be bad for the general public seemed to only be voiced on obscure websites. And while this merger was only the first in a series that has occurred, and will most likely continue to occur, Viacom has been nice enough to stick with only one forte: media control.

Simple media control, however, is not enough for some other companies, as in the case of Vivendi. For Vivendi outright control of natural resources and media outlets is on the agenda.

Your Water is Now in Vivendi's Hands

Within the last ten years Vivendi, along with three other global corporations, has helped privatize water and now helps supply nearly 300 million people in every continent through their enterprise, Vivendi Environment.

Water privatization is usually accomplished with the help of the World Bank, whose loans to struggling and Third World countries often require concessions, such as water privatization. For the World Bank, these are often known as "water supply" loans, and currently out of 276 loans the World Bank has given to governments, water privatization has been a requirement in about one-third of the loans.

Privatization as a condition to government loans from the World Bank has tripled between 1990 and 2002. The World Bank is a development assistance organization aiming to help the poorest people in the poorest countries, while establishing incentives for development in the private sector. Vivendi has developed nicely from it.

Yet while Vivendi privatizes our natural resources, GE destroys them, AOL-Time Warner and Viacom controls what we see and hear and Disney exploits our children, News Corporation takes the cake.

News Corporation: The Business of Censorship

190

While Rupert Murdoch, owner of News Corporation, publicly backed the war efforts of George Bush and Tony Blair in an interview in February, his views on what we should or shouldn't see are much more ominous.

In 1997, in exchange for his Asian-based Star TV, Murdoch agreed to drop the BBC news from broadcast in China, as many of their reports were critical of Beijing. Murdoch has also been responsible for killing the publication of a book critical of China by Chris Patten, and another book critical of Supreme Court Justice Clarence Thomas, in which authors concluded that Anita Hill had told the truth.

After a request from the White House for censorship during wartime, mainly of videos that contained speeches by Osama Bin Laden, Murdoch acquiesced and was quoted as saying, "We'll do whatever is our patriotic duty."

Do I really need to say anymore? What can be more disconcerting than the free press finally doing whatever is their patriotic duty, especially by censoring information? But the fun doesn't stop there, let's look at Henry Kissinger.

Murdoch and Kissinger: Symbiosis as Survival in the Public Eye

In 1997, Henry Kissinger was kind enough to present the humanitarian of the year award to Rupert Murdoch, and while Murdoch receiving a humanitarian award after his track record of censorship in the media is funny, Kissinger giving a humanitarian award is even funnier. If you don't remember Kissinger's crimes against humanity when he was secretary of state, here's a little snippet from Christopher Hitchens, author of "The Trial of Henry Kissinger," which has now been made into a documentary movie:
VIETNAM
Kissinger scuttled peace talks in 1968, paving the way for Richard Nixon's victory in the presidential race. Half the battle deaths in Vietnam took place between 1968 and 1972, not to mention the millions of civilians throughout Indochina who were killed.
CAMBODIA
Kissinger persuaded Nixon to widen the war with massive

bombing of Cambodia and Laos. No one had suggested we go to war with either of these countries. By conservative estimates, the U.S. killed 600,000 civilians in Cambodia and another 350,000 in Laos.

BANGLADESH

Using weapons supplied by the U.S., General Yahya Khan overthrew the democratically elected government and murdered at least half a million civilians in 1971. In the White House, the National Security Council wanted to condemn these actions. Kissinger refused. Amid the killing, Kissinger thanked Khan for his "delicacy and tact."

CHILE

Kissinger helped to plan the 1973 U.S.-backed overthrow of the democratically elected Salvador Allende and the assassination of General René Schneider. Right-wing general Augusto Pinochet then took over. Moderates fled for their lives. Hit men, financed by the CIA, tracked down Allende supporters and killed them. These attacks included the car bombing of Allende's foreign minister, Orlando Letelier, and an aide, Ronni Moffitt, at Sheridan Circle in downtown Washington.

EAST TIMOR

In 1975, President Ford and Secretary of State Kissinger met with Indonesia's corrupt strongman Suharto. Kissinger told reporters the U.S. wouldn't recognize the tiny country of East Timor, which had recently won independence from the Portuguese. Within hours Suharto launched an invasion, killing, by some estimates, 200,000 civilians.

One might think that Kissinger is the perfect man to give away a humanitarian award; after all, he did win the Nobel peace prize in 1973 along with Le Duc Tho, peace talk spokesperson for the Communist Party of Vietnam in 1973. Le Duc Tho turned down his prize saying that hope for peace was premature, while many were saddened to see that Pol Pot, Suharto and Pinochet could not make the award ceremony for Kissinger.

Kissinger may still face extradition to Chile for questioning regarding the overthrow, but his luck has held out after narrowly escaping the clutches of French and Spanish courts for the same crime. Along with numerous lawsuits and accusations well chronicled and documented in everything from the "New York Times" to the "Irish Times" (while surprisingly absent from Murdoch news outlets), Kissinger appears to be a man on the run with only the international community attempting to hold him responsible for his crimes.

Yet the irony continues. Two months after the attacks on the World Trade Center and the Pentagon, President Bush named Henry Kissinger to an independent panel to investigate U.S. intelligence failures that may have led to September 11.

Appropriating "Ground Zero"

As a result of those airplanes-turned-cruise-missiles, the American flag has been flown furiously in some sort of hope that the flag might save us. The news media was quick to connect two skyscrapers and a five-sided building with Old Glory and ideas of patriotism.

While the victims played the major role on television, they were exploited in the most sickening way as the term "ground zero" began to hit the airwaves. Even now if you run the words "ground zero" through the Internet, any reference to Hiroshima, Nagasaki, or any sort of nuclear test has been totally buried under references to September 11. To equate the pain and suffering of the survivors and family members of September 11 to the pain of using nuclear weapons on civilian populations is very disturbing.

Even today after the nuclear bombing of Hiroshima and Nagasaki, which resulted in an estimated death toll of up to 340,000 people, Americans still feel the need to exploit and debase the memories of other victims to make theirs feel more tragic, more heartfelt, more real and more isolated. In 1945, 340,000 civilians died in Japan because of America, and Americans have the audacity to compare themselves to the pain and suffering of those people.

What is more revolting is that seven months prior to the world entering the nuclear age, President Roosevelt received a 40-page report from General Douglas MacArthur containing five separate steps the Japanese would take to surrender. In other words, the use of nuclear weapons was not necessary. The steps were nearly identical to those the Japanese accepted on September 2, 1945, after the subsequent nuclear holocaust its people experienced.

Why did the American public not know this until a newspaper article was released about it ten days after Nagasaki? The answer: Self-imposed wartime censorship. Reporter Walter Trohan of the Chicago Tribune obtained the memorandum from Presidential Chief of Staff Admiral William D. Leahy, and while some have questioned the authenticity of the memorandum, neither the White House or State Department ever challenged the article, and General MacArthur later confirmed the article's accuracy.

The Media: Watchdogs or Cheerleaders?

So what role does the media play? Historically, it has been the cheerleader for big business and government. And being that the two seem so inexplicably intertwined, it seems that it has only been invested in the interests of the elite.

The same people to whom the President gave a $1.3 billion tax cut are the same people that exploit their workers, making the rest of the population (the other 99%) shoulder what they don't have to pay. These people are the ones that pick what you watch, what information you absorb, and in the long run, most likely make our foreign policy decisions.

From the global policy forum, a group which monitors the United Nations Security Council, the following statement on Iraq was written around 1999:

Iraq has the world's second largest proven oil reserves. According to oil industry experts, new exploration will probably raise Iraq's reserves to 2-300 billion barrels of high-grade crude, extraordinarily cheap to produce, leading to a gold-rush of profits for international oil firms in a post-Saddam setting. The four giant firms located in the US and the UK have been keen to get back into Iraq, from which they were excluded with the nationalization of 1972. They face companies from France, Russia, China, Japan and elsewhere, who already have major concessions. But in a post-war military government, imposed by Washington, the US-UK companies expect to overcome their rivals and gain the most lucrative oil deals that will be worth hundreds of billions, even trillions of dollars in profits in the coming decades.

BIA – Bureau of Iraqi Affairs?

Possibly this group is wrong. If they were right, they would be on television for sure, right? But they are most likely wrong. Colin Powell said the U.S. would hold Iraqi oil "in trust" for the Iraqi people. Note the quotation marks around "in trust." I have not been able to locate one news article that did not have these words in quotation marks, let alone a description of what this sort of trust is.

As it is, if oil in Iraq will be used for the benefit of Iraqi people, what about the oil in Alaska, Oklahoma, Texas and other places? I still haven't received any sort of benefit from oil extraction that has taken place in this country. Regardless, if the Iraqi "trust" is anything like the Indian trust funds initiated by the Bureau of Indian Affairs, you know it won't ever make it to the Iraqi people.

Maybe you missed that story, too, though. Remember when the Secretary of Interior couldn't find nearly $100 billion that was supposed to be held "in trust" for Native people?
So what role does the media play? And why are we still listening?

BLOOD PAGES
Sara Marie Ortiz

Part I.

I am afraid as I am writing this. I am afraid of what may emerge. I am afraid that the words won't come out right and, that when you read this, you will not understand me.

I come from a place of darkness. I was told to look to the silver lining, "Over there, see that cloud on the horizon, isn't it beautiful? Now, doesn't it make everything okay Sara?" I have lived these days of my life not knowing. And now, that I am closer than ever to knowing... I don't want to know anymore. I want to forget everything. I don't want to write it down, or take a picture. I don't want to save these moments filled with a mother's neglect, homicide, rape and addiction that betrays logic to my hard drive. "The world is yours", they said. They lied.

I come from a desert region, deserted by purity and simple shame long ago. Albuquerque, New Mexico. "Land of Enchantment" or "Land of Entrapment", as it is called by some of the less modest locals. Most of the white people packed their things and had fled by the early nineties, the small families of them, from my tiny urban Westside development. Sky View West, and the closest liquor store was a whole mile and a half away when I was born. But, of course, it's not like that anymore.

Sesame Street was my garden. I was delivered by a doctor on the Southside. I was pulled from my mother by metal forceps, sterile and sharply unwelcoming on my slippery pink skin. I was pulled, just to be dropped in a playpen full with crack houses, strip clubs, Mexican restaurants and liquor stores. I was pulled, just, to be pushed into the street I was to learn my name on, Sesame Street. How do you get there? Don't ask me. I'm trying to forget. A white doctor brought me into this world in a Southside hospital delivery room, but my mother was the one who fed me to the coyotes.

Part II.

I hold my memories in my palm and turn them over and over, curious, in my hand, like shiny things coated in dust. I would be lost without the snapshots. I am two, maybe three. I am sitting on a plastic lawn chair, brown legs spread in a burgundy jumper made of corduroy, a white shirt beneath it. A ball, red and yellow, suspended in mid-air above me. I can, almost, hear the tiny joyous sounds my throat made then. My mouth is open, my bright smile flashes in the early morning sun. My grandmother and great-aunt, sit on either side of me, light graying hair coiffed, pantsuits pressed. Both of them laughing, my great-aunt's hands are together in front of her, her glee frozen in time, her head is thrown slightly back in amusement.

We are all there in that picture, a backdrop of adobe behind us. My backyard overlooks the Sandia mountains, but only our rosebush is visible in the picture, small red blooms frame the space near my grandmother's face. My aunt is in an assisted living facility in Lubbock now. She was diagnosed with Alzheimers and Parkinins disease in early nine-teen ninety four. She used to work for the FBI in the sixties and early seventies. My mother says she wasn't a particularly generous woman, but she treated my mother well when she was younger, before she married a black man.

The last time I saw my aunt was the summer I turned fourteen. I had been sent to "Peanut Country" to stay with my grandmother in eastern New Mexico as punishment for rebelling against my mother, by lashing out at her for marrying a half Chicano half Navajo hustler/gangster/heroine addict, without telling me. I had begun to stay out later and later. This is when she sent me away.

I listened closely to the conversations my grandmother tried to have with her sister, my aunt, who we saw fading from us before our eyes. My grandmother, my aunt, and myself at the Tastee Freeze, plates of the night's special in front of us. I was not smiling that evening, nor do I smile as I remember it.

My grandmother talked, in her careful way, of people and happenings that I had only heard about and saw in the photo albums she kept on the glass table next to the neatly made bed in the guest room. A thin coating of dust.

I watched my aunt's eyes the closest as my grandmother talked, urging her to recall the places and people she spoke of. My aunt sat rocking slightly, staring at something, seemingly a million miles away, in the distance. I tried to talk to her too, but, with no real success. I wanted to make her remember me, despite the fact that I was told she never would. I was fourteen then, I returned to my neighborhood at the end of the summer, and

197

now only think of her when my family brings up "her condition" when we visit at holidays and special occasions.

Part III.

Bad tastes rise up in my throat, and even the Pepsi can't wash them away. I remember eating oysters for the first time. I am ten. We are all sitting on the patio at The Pelican grill and bar. The evening is clear for the most part, but some clouds over the eastern horizon are moving closer and a cool breeze blows the white linen tablecloth. It is late summer, but it is the desert, so we're dressed light. My blonde lawyer mother, my Indian poet father, with graying black hair, and me.

My father says, "You've never had oysters before Sara?"
I say, "I've never wanted to."

He hands me the cold shell with the jiggly and wet opaque creature, resting on the iridescent surface. I look into the shiny wetness of the thing in my hesitant hand. I close my eyes and raise it to my lips. I do not taste the small oyster, I only feel the cool smoothness of its slippery surface as it slides over my warm tongue. It is only a small wet globe in my throat, and only for a moment. I am changed. I have tasted a creature not coaxed, but stolen from the vastness of the ocean. I poured my soda down my throat to rid my mouth of the remnants of its body on my tongue. I have been offered, but I have never tasted an oyster again.

Part IV.

2002. Not much has changed. The coyotes cry through the mists of the river and swamps near my home. Camino Torcido: criminal road. She brought me here, so I could hear their cries more clearly.

When the drug and gang related crimes started to increase and the Mexican families started calling every house, on the block, theirs, she
lost her job and moved to Santa Fe to work for the public defender. My mother, she left, as she had in my most vivid and shadowed dream. She took my brother and moved to a tourist's city only sixty miles to the north of the hard and crumbling soil that she dreamed, desert mecca dreams, of growing

me from so long ago. The same soil that has bore so many coke addicts and young brown prostitutes and hustlers that I've forgotten to count.

"You need to get out of there, Sar'", she would tell me over a cracking cell phone line.

While driving in the south valley, returning from an unsatisfying deal, a bullet ricocheted off of the driver's side of my mint green Honda Civic. We were stopped at a red light at a busy intersection across from a grocery store, my baby, myself, and her dad, Antonio. I sat still, listening for more sounds. Antonio ducked, instinctively. A car's tires squealing in the night. We pulled into a grocery store parking lot to examine our car. He knew we had been hit by something. I still wanted to believe that the sounds I had heard were those of a car back-firing or a tire blowing out. There was only a small indentation where the bullet had hit, only some of the paint chipping off. As we drove away from the parking lot all I remember was Antonio's speculations, "it was probably a small gun. Maybe a 22.".

All I remember was him saying things, far away, though he sat right next to me, and the sound of my breath slow and broken in my mouth.

I am here, and then I am not. I am sitting in the waiting room of the Planned Parenthood. This wasn't planned. If this is what parenthood looks like, I want none. I flip through pages of tabloids and glossy paged magazines on the table as I wait for the first round of pills and shots.

I am disinterested in the skinny white women and their expensive clothes, the articles about these women and their "problems". Seems every other article deals with ways that the reader can live better.

I am fifteen. I am a mother. I am brown. These articles are black words colored by bright advertisements on shiny white pages that sing empty songs to my hungry and tired eyes. But this doesn't stop me from picking up the Marie Claire at the bottom of the stack. I read about women, because after all, I am one. The girls in these glossy pictures are the color of clay. They are ebony and mahogany. They are thin, and crying out to me, from the flatness of the pages. And the pages are not flat anymore.

BLOOD IS ALWAYS THICKER
Penina Partsch

There are two things that Samoans are best known for, their large size and their *bad* tempers. If you're Samoan and you don't have one, you undoubtedly have the other. My father falls quite comfortably into that second category. Although he's rather small in size, he makes up for it with tempest-sized temper tantrums. His quick to anger personality was usually to the detriment of our entire family, but every now and then it would deliver the greatest excitement. Add the fact that my father was a rock and roll musician, the stereotypical thrasher, for most of his youth and you get one angry Samoan.

It was 1991 and I was twelve, and this summer would mark a monumental occasion for our entire family. We were going on vacation. This may seem like nothing to most people, but our household was composed of two parents, seven kids, a grandmother, an aunt and two cousins. Organizing, let alone finding the money, made family vacations almost impossible and yet, we were doing it. The plan was to leave Houston, hit Dallas, Salt Lake City, and finally Los Angeles, where my cousin whom I had never met was getting married, and then back home. You can imagine the anticipation of seven kids who had never been on such a journey.

The preparation was actually the hardest part. We had no idea how to prepare, and to make things worse, my family was cursed with extreme procrastination. Never mind that we knew about this trip for months, we all decided to prepare the day that we would be leaving. As we rushed to wash clothes, argue over lost shoes, misplace matching socks and accessories, my father's patience began to run short. Despite the fact that he too was a procrastinator, he was a simple man and required only one small bag of possessions which he could pack in five minutes. If my father couldn't find something, then he left it behind without a second thought. No one else possessed that gift. However, my father was running the show and coming close to calling things off. With a simple threat, "Whatever is not in the van in five minutes stays behind, including people," we stuffed whatever we could grab into bags and jumped into the van. Even my grandmother became frantic about being left behind. She zipped up her bags and rushed out the door. What my father said, he meant, and none of us were about to miss out on this trip of a lifetime.

Finally, after a whole day's work and excessive complaining, we finally headed out. Only a few miles into the trip, I began a fight with my Aunt over a change in plans which she had suggested. We were taking Dallas off the list of places to hit and I was just not having it. This was actually a usual occurrence. If I was not fighting with my Aunt over where we would be going, I would have found beef with someone else over one thing or another. My mom asked me, as always, "Why do you have to ruin everything? Just like your father! You've got that bad Samoan blood." Then she'd shake her head and sadly turn away. I could almost always hear her heart breaking over the fact that of all the things I could have inherited from my father, it was his temper. Not his hair, not his eyes, not his prodigious guitar skills, but his temper. My father did not step in, this was not his fight. We seemed to take turns with our tantrums. I suppose we had this silent understanding that should we attack the family together at once, they would leave us. Instead, he told me to calm down, turned up his Santana tape and made us sing along. Oddly enough, he was both the instigator and the peacemaker.

We made it all the way up to Utah without a single word of confliction from my father. Not even a hint of negativity, only cheerful singing and lots of bad jokes. These were my favorite moments with my father. Unfortunately, good times also held an ominous foreshadow. We learned long ago that the longer the calm, the greater the storm. I always cherished the good times with my father, but the fear of his temper breaking free was always in the back of my head.

In Utah, we stayed with an Uncle in Salt Lake City whom we had not seen in many years. For two days we toured the city, historical sites, wax museums and drove through the mountains. Perhaps it was the amazing weather, but my father, in all of that time, did not get upset once. As day two ended and day three began, we loaded back into the van and headed for our next destination: Los Angeles.

I will always remember that drive from Utah to California. It's when I fell in love with the land. I admit I slept most of the way, but when I would stir from slumber, I was surrounded by the most breathtaking views of the countryside. Growing up in the big concrete prison that Houston is, it's hard to appreciate nature. Even the city lights that glimmered far off in the distance had a certain brilliance that I had never before experienced. My father did most of the driving and I would wake to chit chat with him alone about nothing at all. I still smile thinking back on that drive.

When we finally hit Los Angeles, it was daytime and despite

201

our exhaustion from the driving, not one of us could even think of sleeping. The mystery and allure of this new city was too intriguing. My parents could see the anxiousness glowing in our eyes, but were too tired to sightsee. We were not disappointed, however, because meeting this strange, alien family was excitement enough.

The kids in the family were thugs and gangsters. They had grown up in the ghettos of Compton and Long Beach. They were the most fascinating people I had met in my twelve years on earth. I tried my best to make friends with them, but my father kept a watchful eye. He didn't approve of me learning gang signs and getting on the inside of thug mentality. However, being in a new environment, with loved family members he hadn't seen in years, anger was far from his mind. I could have stayed up all night listening to my cousins' stories of guns and 5-0, So Cal slang and all, but my father eventually gave us his familiar nod that meant, "No more." I went to sleep with disappointment that was soon overpowered by anticipation for the next day.

The morning brought great eagerness and my parents agreed to drive us around the city. My father's good mood was beginning to worry all of us, but we decided to take the good times that we could and drove into the city. My poor grandmother fought to see out the window, but all seven kids and both cousins had every window blocked. We kept our faces plastered to the windows, big grins frozen, as we searched for movie stars and awed at locations we recognized from movies. My father drove as he and my mother told stories about when they once lived there.

They both lived in L.A. at different times, before they had met. My father was apparently a bar attraction in his day. With lit eyes, he began by gloating about grinding glass from beer bottles with his teeth. I was amazed that his eyes glimmered even more as he said phrases like, "Sometimes my gums would bleed and everyone would freak out. It was so great!" or "Can you imagine how much glass I must have ingested?" My mother, a complete opposite of my father, simply showed us the bank that was built where her home once stood. She also told of the "cherry" or "choice" cars that she and her girlfriends would cruise in on the weekend. My grandmother chimed in now and then, "Oh yeah, I remember that!"

Strange stories and all, and yet not one grin had broken our faces. We were in a dream. Unfortunately, that dream was about to take a strange turn. The storm was coming, but not like we would have ever anticipated. Suddenly we heard a car horn from behind. My father had cut off another car in traffic. He turned to my mother, who was giving him that motherly look of disapproval, and then shrugged his shoulders. She just shook her head at the familiarity of my father's lack of concern and we rolled on.

Four whole blocks down, the car my father had cut off caught up and pulled up to our van. We could not see a driver from our windows, but no one missed the middle finger that pushed its way intently towards my father, whose fingers now gripped the steering wheel tighter. He was angry. I didn't need to see his face to know that his brow was furrowed over glaring eyes. Then the other car careened closely to us as if trying to drive us off of the road. My father swerved accordingly to avoid a collision and then asked, "Are you kids okay?" and to my mother, "How about you? Are you okay?" Most of the replies included the word yes. My grandmother was a different story.

The swerve had caused her to hit the seat in front of her and her glasses had cut into her forehead. The engine revved loudly and we knew the storm had hit. The rage we had anticipated was at last here, but we were not afraid because this vengeance was not focused at us. Instead, we hopped on the adrenaline rush and aimed our stink-eye at the mystery driver. Even my mother, my calm and understanding mother, yelled into the air, "That Bitch!" My mother only swore when she was very, *very* angry. We were all very angry, but no one could wreak the havoc that we knew my father would. We caught up with the car and attempted to sideswipe it as well. Suddenly there was a red light and both cars came to a screeching halt.

I remember stories my parents retell of my father's notorious temper. They involved an array of shot guns pulled over stolen parking spaces, broken beer bottle fights over disrespectful remarks, broken instruments over fellow band members getting off key, whole homes destroyed because my *poor* father had to wait at a front door too long or bloodied faces and broken bones over the wrong look, just to name a few. I could feel the anger radiating off of my father and I knew that this moment would go down with the long list of quick-tempered fights my father had going.

At the light, the driver leaned over and began to shout. It was an overweight woman in her late thirties. We couldn't hear her angry words through the closed windows, so my father rolled his down. As she leaned closer to her opened passenger door window to give my father a piece of her mind, he did the unthinkable. This was not the usual angry, malicious, destroy-everything-in-sight kind of guy we were used to. He was now the greatest comedian we had ever seen. Just as the light changed to green, my father grabbed his soda pop that had been sitting in the cup holder and threw the contents into the woman's car, all over her front seats and all over her. My father quickly stepped on the gas and roared off.

We stayed in silence only for a moment until we broke into hysterics. The laughter did not last for long, however, because the woman caught up again. Again we came to a red light. As we stopped, she began to sift through her belongings in the back seat of her car. Everyone in our big van froze, petrified.

We were in Los Angeles. This was the city that made it into the news for only three reasons: movie stars, the Lakers and guns. We nervously watched as she produced something metal. We could not see because she moved so fast. She flung open her door and stood, turning to glare at my father. Then she made her way around to his door and held up her weapon. I envisioned a traumatic end to our family vacation, another tragic story from the streets of L.A. Then at last, her hand came into clear view and we saw her treacherous doomsday machine. It was…a colander.

A colander!? Who, or rather *what* was this woman? Who threatens someone with a colander? Of all the things that she could have rummaged out of her car, she chose this. I wondered why in the world she carried a colander in her car. Was she headed to a cooking class? I pictured her vividly recounting this incident to her fellow, pudgy cook mates as they measured ingredients for banana rum cream cake. Perhaps she was a scientist and this was the last piece to her mind control helmet. Regardless of the reason, she chose a colander.

She quickly began to bash the driver's side door with her handy kitchen utensil. My father had had enough. The light changed again and this time he decided to be physical. All of his "bad Samoan blood" must have rushed to his head because with great force, he opened his door, hitting the lady and throwing her into her own car so hard that she bounced and fell to the street. With his temper raging, he sped off leaving her lying there. We watched as she became a speck behind us. No one said a thing. We all wanted to laugh loudly, but we knew he was not calm yet.

We waited quietly, my brothers and sisters and I stifling unstoppable giggles between each other. This woman could have been run over for all we knew, but watching my father in action, that was all we could think about. After a few turns, I suppose when he felt the coast was clear, my father turned around to us and smiled. He knew that we were just waiting for a sign. The storm had passed. My mother was not angry, only in shock, which soon began to fade. She was the first to laugh. "What a weird family we must be," was all I could think as I laughed along with them.

I often sift through these stories of chaotic incidents and ponder the angry temper which I was lucky enough to inherit. I do realize the difference between my father's temper with us and outsiders. When it comes to family, the justification does not necessarily have to be clear. With seven kids under the age of twenty, I give my parents allowances. However, when it comes to my father's temper with outsiders, most often it involves one of his family members, or some other loved one to be threatened. His quick temper is an umbrella that covers more than I had ever realized. At the slightest chance of an outsider hurting one of his loved ones, my father does not hesitate to protect. Had my grandmother not been injured, I wonder if my father would have pursued to the same degree.

Today I fight my grandmother's battles always. Should she call me out, we will argue together. But if someone becomes a threat to her, right or wrong, I am by her side, her defender. I know for certain that I am like my father. My quick temper is often the punch line of jokes at a dinner table, but when needs be, I will fight. No, I didn't inherit my father's great talents or any of his better physical features. I have his temper, but that means that I've also inherited something I'm proud of, his belief that blood is thicker than water, that family comes first and to defend those that you love no matter the consequences.

So yes, I am Samoan and I have a bad temper.

TUSIMAILELAGI: A MESSAGE FROM HEAVEN
Penina Partsch

In March of 1984, my mother was pregnant with her third child. As my mother tells the story, one night she awoke to a sound and began scanning the room, looking for the culprit. The bedroom door was open and through sleepy eyes, she saw a little girl peeking ever so slyly in from the hallway. "Who is that?" my mother whispered out into the stillness of the room. The little girl stepped into the doorway, stood there for a brief moment and then ran away. My mother searched the entire house frantically. She found nothing.

Three months later my mother gave birth to a little girl who she named Tusimailelagi Faustine Moe Partsch. We called her Tusi for short. A year later, we found my mother crying, holding Tusi tightly to her chest. At just one year old, Tusi's distinguishing marks were already evident. Her tiny body was topped with a big round head and perfectly round face, all framed by long, curly tresses. Her smile was a little crooked, curving higher on the right side and her eyes were big and glossy, trimmed with lush lashes. It was this that caused my mother to cry. It turned out that Tusi was the splitting image of the mysterious little girl who visited my mother a year and three months earlier.

Tusi's full name translated from Samoan meant a message from heaven. This was also my grandmother's name. We all knew that Tusi was special. Even I knew this at five years old; however it was not something that I was ready to embrace. For me it was a threat. Perhaps I wished that it had been me, but regardless, she was the one. She carried a grace and a glow that I just did not have. Rather than praise her for it, I held it against her.

Through the next few years, she was just my little sister. The five year difference was a big one, and I was not a good enough older sister to see her as anything more than annoying. She had her friends and I had mine. Whenever there was intermingling, it would always end with a fight. It was your basic older and *supposedly* more "mature" sister vs. the younger and *supposedly* more "immature" sister kind of stuff. This was the consistent friction, but there were other factors too. The things that were not consistent seemed to be the most detrimental.

When we were five and ten years old, my grandparents came from New Zealand to visit for Christmas. They brought so many presents that outshined the twinkling lights and glittering ornaments hanging only slightly above. As most young children do, I searched for my own name amongst the tags on the brightly colored packages. By this time, we had another sister, Pele, only four years old. Her name graced a few presents. Our twelve-year-old sister, being the firstborn had a good number of gifts as well. Tusi, being my grandmother's namesake, had the most. As I searched with youthful Christmas excitement, I was jarred with sadness when I realized that there were none for me. When I brought this to my mother's attention, she just smiled and told me, "Oh, they didn't forget you. In fact here, these three are yours, they just put Tusi's name on there by accident." I took the presents, but I knew that they were not for me. This incident increased my jealousy over Tusi, jealousy that would last quite some time.

Eventually, over the years, the immature bickering Tusi and I would share turned into intense arguments. I have yet to meet anyone who can make me as angry as Tusi can. She had this way of making snide remarks that got under the skin and stayed there. No catchy coined phrases, but in any argument, she could sum things up with a finality that allowed her to win regardless of whether she was wrong or not. Tusi also had the ability to convince my parents that she was always the victim. Tusi, after all was the special one. She was the kind hearted, loving one. Why on earth would anything be *her* fault? *I* was the angry one, the second child, the one who felt they must fight in order to be seen. She was never intentionally sneaky, but if things came down to it, and I was beating her, she would find a way. She would always win. So regardless what the situation was, Tusi was always the victor. This kept me infuriated for years. As time passed, somewhere between the jealousy and the continual defeat, I almost hated her.

In the summer of 1992, my family decided to spend a weekend two hours from our Houston home. We drove to Livingston, Texas, to get away from the city and spend a little quality family time. The entire trip up, Tusi and I fought. We argued like we had never argued before, and all over petty incidents. "Mom, Tusi has my walkman. She won't give it back!" "Mom, she won't stop singing when I'm trying to hear this song!" "Mom, Tusi told me to shut up!" "Mom! She keeps hitting me!" I've never been scolded so many times in such a short period of time.

We finally arrived in the late evening and decided to grab a quick bite at the local Burger King before retiring to a hotel. As soon as we entered the fast food joint, heat consumed us, like putting on a fur coat in the desert. It turns out that their air conditioning unit was out. To make things worse, they had no cold drinks because both their soda and ice machines were out as well. So, we decided to go to McDonald's instead. Their air conditioner was running strong, as were their soda and ice machines, so we stayed there. It took us a little while to organize the food order, but we eventually got our food and took a seat. Then my mother asked for our attention, "Where's Tusi? I haven't seen her for a while."

My father knew, "She's outside on the playground."

"Well, tell her to come get her food," my mom told me.

Tusi was the last person I cared about at all, let alone getting any nourishment, but I was already on my mother's bad side, so I went. Once outside, I saw no one. I came back in and announced that she was not out there. So, I was then instructed to check the bathrooms. Not finding her there, everyone joined in the search. We checked the van, the outskirts of the restaurant and yet, no Tusi. My parents began to panic as an elderly Mexican man approached us.

"Is your daughter about this tall?" he asked slowly showing the approximate height with his left hand, right hand leaning on a cane.

"Yes," my mother said frantically.

"Does she have long, dark, curly hair?" he asked again.

"Yes," my mother cried, "have you seen her?"

"I think so," the man answered.

"Where did you see her? Please tell me," my mother begged, eyes wide.

"Around there," he said pointing to the back of the building, "I saw a girl who looked like that get in a car with some men."

"When?" my mother asked with fear in her voice.

"About fifteen minutes ago."

Then the whole room seemed to turn grey, like a cinematography shot. It was cold too. Cold and grey is what I remember about that moment, then an onrush of tears. I broke down hard. I, the one who despised her, the one who fought with her constantly, even hated her. There I was wailing, shaking from the inside out. The look on my parents' faces made it even harder for me to control myself. There is nothing more painful than watching the face of a mother who has just lost her child. The family just held each other and we cried. No one spoke. Only the sound of wailing echoed off the concrete parking lot and out into the darkness.

I couldn't believe it. She was gone. All of our fighting, all of our bickering, all of the anger ran on fast forward in my head. I hated her so much for so long and now she would never be coming back, ever. I pictured her round face, big eyes squinting because of her crooked smile. I thought of the time I lent her two dollars to buy a gumball machine and remembered how happy she was. How she wouldn't stop hugging me no matter how much I pushed her away. I remembered finding her crying in the dark when my mother had her third child, sad because she would no longer be the baby. I remember comforting her, comforting her and taking care of her like I should have been doing all this time. Then I pictured her little body trapped in the backseat of a stranger's car. I envisioned her crying, screaming, begging for us, begging for me. She was asking why I hated her. I fought against these thoughts. I fought even harder against thoughts of what these strangers might be doing to her. My mind was whirling. So many thoughts, so many fears, so many reasons to cry and that's just what I did. I cried and cried, hanging onto whoever I could reach.

Then my father broke free and ran outside, around the side of the building. That's when it hit my mother as well. We were wasting time. There was no reason to give up now. There was still time to find my sister. Despite the description given by the elderly man, my mother was determined to prove him wrong or at least find these men, these vicious animals that stole her precious little baby. I don't know if it was denial or pure motherly instinct, but she ran and got in the van. My older sister and I had to see where she was going; we had to help find her too, so we jumped in the van with her.

"We'll be right back," my mother shouted from the car as she sped off, not allowing any questions.

We circled the block. Then we circled the area. As we were driving back by the Burger King we had been in earlier, my mother said, "You know, I have a feeling..." So we pulled into the Burger King and ran inside. The girl behind the counter looked at us as if she knew us and I guess she did. With relief on her face, she quickly said, "Follow me, she's back here." We followed the girl back through the kitchen. We entered a tiny office and there she was, fragile little Tusi. Her large head was bent forward as tears fell from her face like a trickling faucet. Then she raised her head and that large crooked smile stretched across her round face. She ran and folded into my mother's breast. They stood and rocked and cried. My older sister and I hugged as well, crying along with them.

Then the story was explained. When we were at Burger King, Tusi asked my father if she could go out on the playground. He said yes and she went. Then we decided to leave and forgot to do a head count. As we were driving out, Tusi saw the van and ran into the restaurant screaming, "Someone stole our van!" Once at McDonald's, my father thought Tusi had asked to go to the playground there, so didn't think too much of her being missing. When poor little Tusi realized that we were no longer in the restaurant, she broke down. That's when the ladies behind the counter took her to the backroom to see if she could remember any phone numbers or addresses. They fed her kids' meals and let her watch the television. It was nice to know that those things weren't enough to comfort her about being separated.

We all got in the car and met back up with the rest of my family. Despite the fact that the last hour was filled with fear and raging emotion, it was perhaps one of the greatest nights of my life.

My mother did have to make a crack at me for being the first one to cry though, "See, always the first to fight and now the first to cry. You better be more grateful from now on!" She was right. I began to wonder now why I *had* been the first to cry. The answer didn't come to me all at once; in fact, it took quite a while. In the meantime, I was much easier going with Tusi, not so quick to get mad, and we became closer.

What I did eventually realize is that I loved her. All of our fights had in some ways made us closer than everyone else. We had a different bond. To this day she is one of the only people to whom I can tell everything and expect an honest response. I wish I had spent those years appreciating her intelligence instead of resenting it. All of the jealousy and all of hate was finally shattered by exactly what she was; a message from heaven.

SEEING LETTERS
Orlando White

Everything I spell out requires this: Alphabet.

It was a notion I did not know when I was six years old.
In kindergarten I was more interested in the images displayed next to
each letter on flash cards. But it was the letters, which confused me.

I recall my mother playing a word puzzle. She'd circle a line of letters
amongst many other letters scattered on the page. She treated each
word carefully without the pen touching it. Then, she would give me
the pen. I circled random letters. She'd smile and give me a hug.

My mother once told me that my step-dad found a picture of my real
father. He ripped it up. I still do not know who my father is to this
day.

I always called my step-dad, David. And he called me by my middle
name. To him it was better than looking at me and calling me "son."
I am still ashamed of my middle name.

He tried to teach me how to spell.

I showed him homework from my first grade class. It was a list of
words assigned for me to spell out. He looked at me as he was
sharpening a pencil with his knife. I remember the way he forced my
hand to write. How the pencil stabbed each letter, the lead smearing.
I imagined each word bruising when I stared at them.

The words reminded me of the word puzzle.

But, without images it did not mean anything at all.

He said, "Spell them out." I could not.
"Then sound them out first!"

I recall a day, like many other days in grammar school when an older boy made fun of me because I could not speak proper English. I always mispronounced words and I would wonder how to spell them.

I still could not move the pencil in my hand. I saw the letters lined up on paper, but I wanted to circle them.

He then shouted out, "Spell them out you little fucker! I am going to hit you if you don't."

I remember the shape of his fist.

It happened. No one was around, not even my mother. It was as close to intimacy I got from my step-dad. I did not say anything to anyone. He bought me toys as an act of confession. I forgave him.

When David hit me in the head, I saw stars in the shape of the Alphabet. Twenty years later, I found that my fascination for letters resulted in poems.

TRISTAN AHTONE is from the Kiowa Tribe from Oklahoma. He has lived in various points north and east of Santa Fe, New Mexico. He is in his second year enrolled in the Creative Writing/Visual Arts program. He is the Editor of the Institute of American Indian Arts Chronicle. In 2004 he won the American Indian Higher Education Consortium Conference Best of Show, Peoples' Choice Award and Honorable Mention for Digital Art. Other samples of his different works can be found at www.cowboykiller.com.

BRITTA ANDERSSON is a second-year student in the Creative Writing Program.

ISHMAEL M. ANTAR is Gros Ventre from Fort Peck and Fort Belknap, Montana. He was born in Albuquerque, New Mexico. He grew up all around the country and spent time in California, Arizona, Washington D,C., South Dakota, and most recently came back to New Mexico. He's been writing since he was 15 years old and likes to use every opportunity to share his work. He was first published in 1999 in an anthology released by Creative Communication based in Utah. In 2000 he was inducted into the International Society of Poets. He is currently pursuing a BFA in Creative Writing from the Institute of American Indian Arts. freeformlife@hotmail.com

JODY BARNES is Menominee from Wisconsin. She has a ten-year-old son with Autism. She decided to become a writer because she couldn't be a rockstar.

LESLIE GEE is Caddo/Delaware/Choctaw from Oklahoma. She majored in Native American Studies at the University of Oklahoma, Federal Indin' Law & Policy at Oklahoma City University and English Literature at the University of Central Oklahoma before transferring to the God almighty Institute of American Indian Arts where she is pursuing her Bachelors in Creative Writing. She and her two daughters live in Santa Fe.

JAVIER GONZALEZ is from Bogota, Columbia.

JAMIE NATONABAH is Diné from Fort Defiance, Arizona. She is currently working toward her A.F.A. in Creative Writing in Santa Fe,

New Mexico. At the moment, poetry and fiction are at war beneath her skin.

ALICIA NATEWA is from Zuni, New Mexico. She has attended UNM in Gallup, NM and is now attending the Institute of American Indian Arts in Santa Fe, NM to receive her AFA in Studio Arts.

DG NANOUK OKPIK is Inupiat from Alaska. She graduated with an Associates of Liberal Arts in 2002 from Salish Kootenai College on the Flathead Reservation in Northwest Montana. She will graduate with her AA in Creative Writing from the Institute of American Indian Arts in Santa Fe, NM and will continue in the BFA Program. She is an Honor Student, has received the Truman Capote Literary Award, Thayden-Boyd Scholarship, and received the Naropa Summer Writing Program Scholarship Award. She will be published in University of Arizona Red Ink, 2004 edition, and in the University of Kansas, Touchstone, 2004, edition.

SARA MARIE ORTIZ is of the aacquumeh' Hanoh-, the Sky City people, or the Acoma Pueblo people. Ms. Ortiz is a published poet, a graduate of the Institute of American Indian Arts, and is the youngest daughter of poet, professor, laureate Simon J. Ortiz. Ms. Ortiz is pursuing her master's degree in American Indian Studies, and is currently pursuing a career in the development and facilitation of Indian education and new Indian leadership initiatives in Indian communities worldwide. Ms. Ortiz says of her art, and her life work: "This is the story. We are from it and of it. We must pay continual homage to this story we are all from- the story we are all telling with our lives and deaths. We take great care with the story, our story, since it has been caring for us all this time. The children. They are dying. The elders. They are Dying. But there are those among us who live, and who want to do what's right for the story. For the people. I am only one of those people. Fourth world rising. Not just a vision anymore- we're living it now." Feel free. If you'd like to know more, e-mail Sara Marie Ortiz, night or day, at Native Scientist@aol.com.

DELEANA OTHERBULL is Crow/Northern Cheyenne from Montana. She grew up in Portland, Oregon. She has been writing since her freshman year in high school. She will be graduating with her AA in Creative Writing and will continue in the BFA program at the Institute of American Indian Arts. This is her first publication.

PENINATUTASIO PARTSCH is Samoan and her father is High Chief, Savea Partsch. She is in the AA Creative Writing Program and would like to continue in the BFA Creative Writing Program. She is interested in furthering her education in New Zealand, pursuing Film Studies.

HOKA SKENANDORE writes, "My writing is a by-product of reading too much and having uneven levels of chemicals in my brain. I need to write more and sleep less. For further information please contact your local institute for the criminally insane. Machine wash cold, delicate cycle. DO NOT OPEN, may cause serious INJURY or DEATH."

DOUGLAS TWO BULLS is from Rapid City, South Dakota. He is a first-year Creative Writing major.

ORLANDO WHITE is Diné from Sweetwater, Arizona. His clans are of the Zuni Wateredge People and born for the Mexican. He holds an Associate of Arts degree in creative writing from the Institute of American Indian Arts and is currently in the B.F.A. program. He is the recipient of the Truman Capote Writing Fellowship (2002), Naropa Poetry Scholarship (2003), and the Idyllwild Arts Poetry Scholarship (2004). His poems have previously appeared in Red Ink Magazine, Serial Vox Review, Tribal College Journal, and 26.